LITTLE ENGLAND

Plantation Society and Anglo-Barbadian Politics, 1627–1700

Ligon's map of Barbados circa 1640.
SOURCE: Richard Ligon, *A True and Exact History of Barbados* (London, 1657).

LITTLE ENGLAND

Plantation Society and

Anglo-Barbadian Politics, 1627–1700

GARY A. PUCKREIN

NEW YORK UNIVERSITY PRESS
NEW YORK & LONDON
1984

Library of Congress Cataloging in Publication Data

Puckrein, Gary A., 1949–
Little England.

Originally presented as the author's thesis
(doctoral)—Brown University.
Bibliography: p.
Includes index.
1. Plantations—Barbados—History—17th century.
2. Slavery—Barbados—History—17th century.
3. Barbados—Politics and government. I. Title.
HD1471.B35P82 1984 307.7'2 83-19492
ISBN 0-8147-6587-4

Clothbound editions of New York University Press Books are Smyth-sewn and printed on permanent and durable acid-free paper.

To my mother and father

Contents

Illustrations

Tables

Preface

The most eastward island in the Caribbean archipelago is Barbados. It was a British colony from its first settlement in 1627 until 1966. Today, when our thoughts drift to this tiny oasis in the sea, they tend usually to be daydreams of pleasurable calypso polyrhythms and mildly tinted blue waters washing the tension from our perhaps overcivilized bodies. Among specialists, the lively comparative slavery debates have created a ripple of curiosity about the early history of the island, particularly since it was the first successful English slave plantation society in the New World. Recently, several fine studies have appeared that provide much useful information about the colony's development in the seventeenth century.[1] Primarily concerned with answering questions related to social relations and to the productivity of the plantation system, they tend to rely heavily upon pre–World War II scholarship for their understanding of political events.[2]

Written before the advent of the new social history, these earlier political narratives are rather narrowly conceived; the social realities of plantation society are generally treated as if they had no bearing on political matters. Their attention is almost exclusively focused on the planter class. Let us take as a specific example their treatment of the political crisis that engulfed Barbados in the middle of the seventeenth century.

Surrounded by a potentially hostile mob of slaves, indentured servants, and Irish freemen, Barbadian sugar producers stood poised to fight the English nation between 1650 and 1652. The accepted interpretation, based primarily on the research of Nicholas Darnell Davis and Vincent Harlow, portrays this conflict as an extension of English politics. They maintain that Royalist refugees, after the defeat of king and cause, fled the English stage for the less oppressive atmosphere of Barbados. This exodus began sometime in 1645, and by 1650 their numbers had so greatly increased that they were able to forcibly take over the Barbadian government, declare themselves for Charles II, and banish a large portion of the "old planter" class, who were pronounced Parliamentarians.

Writing some fifty years before Harlow, Davis began the tradition of seeing the Barbadian civil disorders as a Caribbean projection of English events. Davis believes that Englishmen who settled in the West Indies were in no way changed by the New World environment. "English who emigrate to the colonies," he writes, "remain Englishmen still; and so it was with the Cavaliers and Roundheads of Little England, as Barbados is boastingly called by her Islanders." The colonists "carried with them even their party spirit," and it was these political preferences that lay behind their activities in Barbados.[3] With the Royalist ranks swelled by retreating monarchists from the mother country, Davis believes that the Cavaliers were soon in a position to take control of the island. Although less explicit about his assumptions than Davis, Vincent Harlow reached virtually the same conclusions about the origins of civil disruption in the colony.

The difficulty with these authors' approach is that it takes no notice of plantation society. It assumes a simple transmigration of political behavior from metropolis to colony. Davis and Harlow are therefore quite satisfied to learn that English factions and parties, along with their ideologies, were transferred and exactly duplicated in Barbados. They are not startled by the discovery that two dissimilar social systems, separated by an ocean and operating in different material conditions, were able to produce the same political divisions. In their account, slaves are invisible, and poor whites are without influence on the actions of the ruling plantocracy. It is not simply a matter of ignoring the average man; they also turn a blind eye to the inner tensions and strains of plantation society that helped to shape the direction of politics in the colony. The effect is to make Barbados appear to have been another English county.

Post–World War II scholarship has gone a long way toward sketching out the daily life of the lower classes in plantation America. But since it continues to depend upon earlier studies for its understanding of political relationships, we are still without an integrated sense of the political history of New World plantation communities. To date, for instance, there is no comparative treatment of the political institutions through which plantation societies were governed. If Barbados is any example, there is good reason to believe that wherever the slave plantation system emerged in the Americas it created closely related social, political, and military problems for the ruling elite.

Before 1627, Barbados was an uninhabited rain forest; twenty-five years later, with nearly 38,000 people living on it, it was a booming plantation society. The men and women responsible for this dramatic change came from vastly different cultural heritages. A majority were African tribesmen, and although Englishmen controlled the colony socially and politically, there were also Irish, Scottish, French, and Dutch settlers on the island. In the seventeenth century Europeans and Africans held markedly different views about the nature of community existence, and they had no prior exposure to the other's style of life.

Pushed by economic dislocation, English society on the eve of colonization was moving from a medieval to a more recognizably modern form of social organization. The former was dominated by a landed nobility whose tenants were, in many instances, also their subjects. It was a patriarchal society in which notions of family, community, and religious devotion were cohesive social agents. By the time Englishmen began settling in the Western Hemisphere in the seventeenth century, the medieval landlord-tenant relationship had evolved into a purely economic association, with the central government assuming many of the civil duties that were once the function of the landed aristocracy. In this latter age, family, community, and religion were still important forces in English life, but there was now a greater willingness to see production for the marketplace and for self-interest as acceptable social concerns. Indeed, dissatisfied with limited economic opportunities in England, many English colonists were drawn to Barbados by visions of self-improvement. Their desire for material success played a central role in the rise of the island's slave plantation system.

It was not a voluntary decision to separate themselves from their native lands that brought Africans to Barbados. They were forcibly removed from their communities and dragged out to the Caribbean frontier. They came from different parts of the African continent, from numerous linguistic and tribal groups, and from different communities in any region. They did not share a common culture or history in the same sense that English colonists shared a common heritage. There were some similarities in their social outlook; by and large, they were from societies where kinship, lineage, and religion were powerful social forces. These elements, however, varied in content and emphasis from one African community to another to such a degree that they could not and did not serve as an immediate bridge over the linguistic and social barriers that divided Africans in Barbados. Ultimately, the heterogeneity of this population facilitated its incorporation into the emerging plantation system.

Englishmen and Africans came from different worlds, but in a remarkably short period of time they built a social order that anyone familiar with traditional English or African societies would have described as being a mixture of the former and the latter but essentially distinct from both. In the colony, vague ideas and expectations gave rise to new social and economic relationships; old taboos became public virtues, and individuals lost one cultural and ethnic identity and acquired another.

Political events in Barbados were affected by the structure of the colony's plantation system—a majority African slave population ruled by a minority European planter class. In studying Anglo-Barbadian relations in the formative years of settlement, the present work seeks to describe the impact the slave plantation system had on the political behavior of the island's governing plantocracy. The central argument is that social conditions in the society had a stabilizing influence on Anglo-Barbadian politics, as planters came to depend upon England to help them resolve some of the problems that were a direct outgrowth of the structural organization of the island's slave plantation system.

These findings suggest that some modifications in our understanding of Anglo-American relations in general might be in order. Traditional wisdom on the subject is based on the writings of Charles M. Andrews. Reacting to the nationalistic works of George Bancroft and Herbert Adams, Andrews argues that no imperial relations in the territorial sense existed between England and her New World possessions, that before 1763 when men spoke of empire they meant "the self-sufficient empire of the mercantilist rather than a thing of territory, centralization, maintenance, and authority." "England's interest in these colonies," he continues elsewhere, "was not political but commercial," and "colonies were not looked upon as the stuff out of which an empire was to be made, but rather as the source of raw materials. . . . No one at this time took any other view." Thus, Andrews concludes that "imperial and imperialism have no place in the vocabulary of our early history."[4] After 1763, however, British policy became more imperial-minded; reacting to this change, Americans came to see themselves as colonials, setting the stage for the American Revolution.

In a recent work of far-reaching significance, Stephen Webb rightly questions Andrews's depiction of Anglo-American relations before 1763. By denying the existence of an imperial connection between England and its American colonies, Andrews in effect writes off America's colonial past. Professor Webb quite properly declares that it is time to "end the commercial and colonial monopoly of interpretation."[5]

By contrast with Andrews, Webb contends that from the beginning English colonization was at least as much military as it was commercial. "The forces fostering imperial attitudes and institutions are clearly visible in the political role of the English army and the administrative careers of its officers' commanding colonies—the governors-general." Their imperial influence was coeval with colonization itself. The empire they organized originated almost two centuries before 1763. He traces its military foundations back to the Tudor monarchies of the sixteenth century. By the Restoration in 1660, paramilitary power relationships were applied over ever widening territories, first in the British Isles, then in English overseas provinces. The essential element, as Professor Webb sees it, was the "imposition of state control on dependent people by force."[6] The result was that by 1681 Anglo-American relations were dominated by an imperial system whose agents were military men ruling in an arbitrary military fashion.

The present study lies closer to Webb's vision of Anglo-American politics than it does to that of Andrews, insofar as it recognizes the existence of English imperialism before 1763. It differs from Webb's view in its understanding of the factors that maintained the colonial relationship. The long and the short of it is that the type of military occupation and intimidation that Webb writes about did not occur in Barbados. England's ability to use force against the colonies must be seen as an instrument of last resort to preserve the imperial connection. Only on one occasion in the seventeenth century did the mother country have to apply naked force to thwart Barbadian desires for home rule. In the main, planters understood that the island's slave plantation system could not survive outside the protective umbrella of the metropolis, and so they accepted imperial rule in exchange for protection.

Taking Barbados as an example, future research might well show that, while other English possessions in the New World were structurally different from Barbados's slave economy, social and demographic forces may have also bound them to the English colonial system. In the eighteenth century the English North American colonies collectively overcame these limitations and successfully challenged imperial rule.

This study is written with several audiences in mind, and perhaps its design should be made explicit. The first five chapters look at the social, political, and economic forces that converged to produce the Barbadian plantation system. The first chapter gives a quick overview of England and Africa in the period immediately preceding the settlement of Barbados. Most students will be well versed in either African or English history, but few will know both. In order to speak to the broadest possible audience, the chapter assumes little familiarity with the subject matter on the part of the reader. The intention is to provide some general bases to measure the extent of social change in seventeenth-century Barbados. To those scholars who have expertise in either or both areas, I apologize for the pace at which this chapter moves; I also wish to beg their indulgence for taking a firm stance on issues that are hotly debated by experts in these areas. To do any more would have slowed the opening chapter to a virtual halt. Chapters 2–5 are specifically concerned with events in Barbados. In considering political matters, I have gone into some details; this is essentially new material that may be of benefit to future research. The last five chapters are concerned with the nexus between the island's slave plantation system and Anglo-Barbadian relations. They were written as the discussion about the value of political narratives was taking shape, and I hope that they have profited from the dialogue.

This is not a pleasant story, as most students of plantation America could anticipate, but when told frankly and openly much can be learned about the evolution of plantation communities and their impact on modern social patterns. It is hoped that this book will remind the heirs of the democratic tradition that their societies were built on social foundations that have to be maintained if their children are to continue to enjoy the fruits of life in a free society.

It is difficult in a few lines to express the sense of gratitude that I have for the many people who, over the years, have contributed to my development as a historian and to the completion of this study. Strangely enough, in this age where knowledge has been overly compartmentalized, it was a philosopher, Arthur Benson, who taught me the wonders of scholarship. As an undergraduate in his class, I came to realize the meaning of scientific inquiry. It was in the history department of Brown University that an untutored curiosity was taught the historian's trade. Particular thanks must go to Perry Curtis, William McLoughlin, R. C. Padden, and Gordon Wood for their thoughtful instructions on the writing of history. In the classroom of Rhett Jones I was encouraged to look beyond the limits of traditional scholarship. Special thanks must go to David Underdown. Patiently, he introduced me to English history, and in the process he made me aware of the importance of meticulous research. In the course of putting together the present study his timely advice saved me from many a fatal blunder. Carl Bridenbaugh graciously made available to me the benefits of his work on the seventeenth-century West Indies. In England, Christopher Hill warmly received a babbling graduate student and offered some useful suggestions. The seminar of Jack Fisher at the Institute of Historical Research (University of London) provided many delightful evenings of thoughtful conversation. A year at the Shelby Cullom Davis Center for Historical Research, Princeton University, where I was on leave from teaching responsibilities, provided me with an uninterrupted block of time in which to transform the dissertation into the present monograph. The weekly Davis seminars were an important intellectual experience, and this book has profited from the discussions. I am especially grateful to Lawrence Stone, Director of the Center, for his helpful comments and suggestions. My colleagues at Rutgers University have been extremely supportive, and special thanks are due to Gerald Grob, Tilden Edelstein, Maurice Lee, Richard Kohn, Robert Gottfried, Allen Howard, and Daniel Horn. In Barbados some exciting hours were spent with Peter Campbell, whose knowledge of early Barbados is extensive. I am especially indebted to Senator Clyde Griffeth of Barbados, who took time from a busy schedule to facilitate my research. I would also like to thank Lawrence and Patricia Streeks for the warm hospitality they showed me during my stays in Barbados.

One of the greatest obstacles to historical research in the seventeenth-century Caribbean is that the data are scattered throughout the archives and libraries of Europe, the West Indies, and North America. Fortunately, my work was made easier by the staffs of various institutions that had custody of documents that were pertinent to this study, and I would like to thank for their assistances: the Institute of Historical Research (University of London); the Public Record Office, England; the British Library; the Royal Commonwealth Society; the Dr. Williams Library; the Bodleian Library (Oxford University); the Guildhall Library; the Hertford County Record Office; the Derby County Record Office; the Society of Friends Library; the Historical Manuscript Commission; the Lambeth Palace Library; and the National Maritime Museum. I am particularly grateful to the Marquis of Lothian, who allowed me to photocopy documents in the Coke Papers. In the same regard, I must thank the Scottish Record Office for providing me with a microfilm copy of the Hay of Haystoun Papers. Ms. Christine Matthews and her staff at the Department of Archives, Barbados, were extremely patient with my many demands. In this country, courtesy of the Edward E. Ayer Collection, the Newberry Library; the New York Public Library; and the Henry E. Huntington Library and Art Gallery generously photocopied valuable material in their collections. The staffs of the John Carter Brown, the Rockefeller, and the John Hay libraries at Brown University and the Alexander Library, Rutgers University, were particularly kind and eager to advance my research.

Grants from the Beneficial Foundation, Brown University, and the National Fellowship Fund supported graduate research in England and Barbados and the writing of the dissertation. The research and writing of the book were made possible by grants from the Southern Fellowship Fund; Shelby Cullom Davis Center, Princeton University; Junior Faculty Fellowship, Rutgers University; Research Grant, Rutgers University; and Social Science Research Council.

All dating and spelling has been modernized, except where it destroyed the meaning or tone of the original citation. The year is regarded as beginning on January 1, not March 25.

Abbreviations

Bar. Archives	Barbados Archives, Lazaretto, St. Michael
C.S.P.C.	W. Noel Sainsbury et al., eds., *Calendar of State Papers, Colonial Series, America and the West Indies*, 43 volumes (London, 1860–)
P.R.O., C.O. 1/1/1, etc.	British Public Record Office, Colonial Series, class 1, vol. 1, p. 1
Hay Papers (S.R.O.)	Hay of Haystoun Papers relating to Barbados, Scottish Record Office
Davis Collection	N. Darnell Davis Papers, Royal Commonwealth Society, London
Dunn, *Sugar and Slaves*	Richard Dunn, *Sugar and Slaves* (North Carolina: University of North Carolina Press, Institute of Early American History and Culture, 1972)
Bridenbaugh, *No Peace*	Carl and Roberta Bridenbaugh, *No Peace Beyond the Line* (New York: Oxford University Press, 1972)

Winthrop Papers	*Winthrop Papers*, 5 vols. (Cambridge, Massachusetts: Massachusetts Historical Society, 1931–19)
Jour. Bar. Mus. Hist. Soc.	*Journal of the Barbados Museum and Historical Society*

LITTLE ENGLAND

Plantation Society and
Anglo-Barbadian Politics, 1627–1700

PART I

CREATION OF A PLANTATION SOCIETY

CHAPTER I

Old World Beginnings

The first recorded visit of an Englishman to Barbados was that of a Captain Gordon, who claims to have landed in the year 1620, but it was another seven years before serious steps were taken to build a settlement. In 1625, Capt. John Powell, the elder, touched at Barbados on his return voyage from Pernambuco, Brazil, to England. Captain Powell was well experienced in sailing the waters of the New World. As early as 1619 the sources have him bringing supplies to the English colonies of Virginia and Bermuda. Indeed, there is reason to believe that he helped to transport the first black slaves to Virginia.[1]

Powell and his party landed on the Caribbean, or leeward, coast of Barbados at a point where the village of Holetown now lies. Amid the rocks and rough surf, there were no safe harbors or landing places on the Atlantic coast. The experienced eye of a seventeenth-century military strategist could well appreciate the advantages of this natural barricade. All attacks mounted against the island had to be made on the Caribbean coast, where defenders could concentrate their forces. As the settlement developed during the century, planters built four towns on the leeward coast: Bridgetown, the island's chief port and present capital; and the three subsidiary hamlets of Speightown,

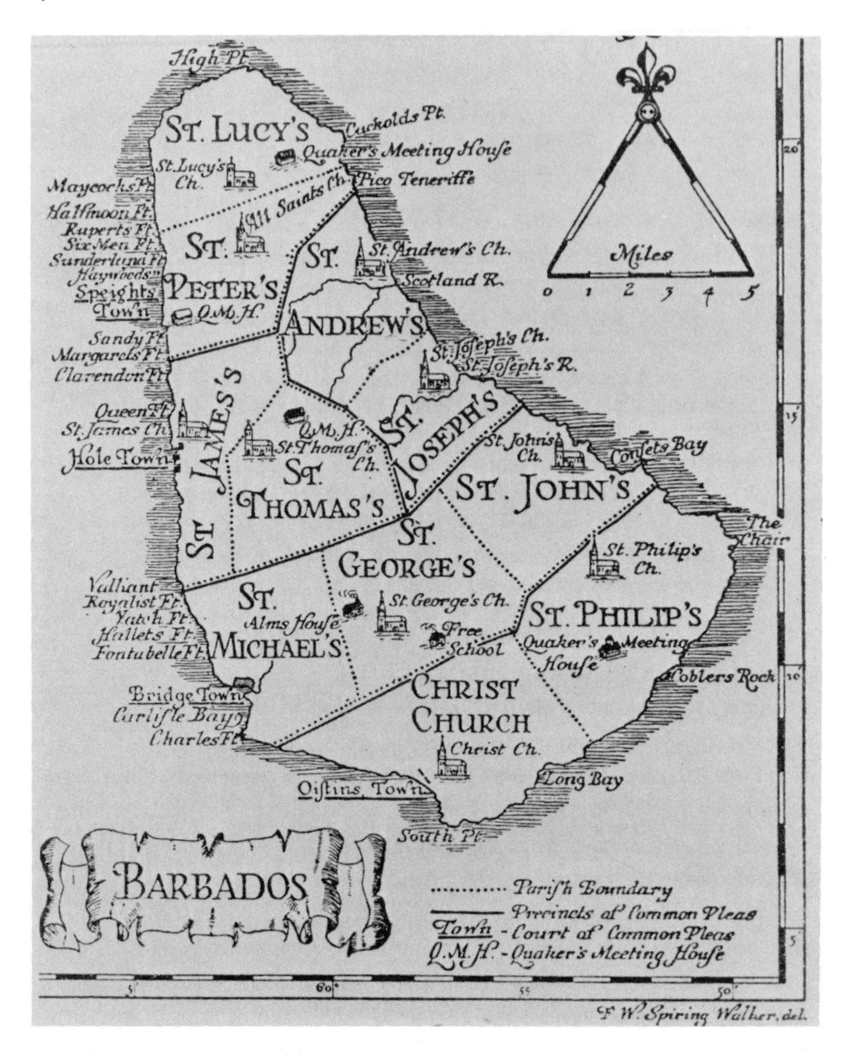

Map of Barbados.
SOURCE: Vincent Harlow, *A History of Barbados, 1625–1685*, p. 335.

Holetown, and Oistins. At the time Powell first explored Barbados, the island was covered by a dense tropical rain forest, which soon convinced the captain that the land was unoccupied. The island was indeed uninhabited, but Arawak Indians coming from South America into the West Indies created a stable agricultural and fishing industry on the island for some centuries. In the sixteenth century the Arawak population disappeared for reasons that have not yet been satisfactorily explained.[2]

Man was absent, but there was life on the island. Barbados was the home of a wide variety of plants and animals. One visitor reported seeing pelicans nesting in trees and turtledoves who "sing us music daily into our ships." Everywhere there were signs of great fertility. The sailors who wandered along the Barbadian coast with Powell feasted on the wild figs, plantains, oranges, pineapples, watermelons, muskmelons, and other fruit that abound on the island. It is easy to understand why Powell might have failed to notice that the island's small size (166 square miles) did not preclude striking variations in weather and topography.

Holetown, the site of Powell's landing, is part of the leeward district that consists of two subregions: the coastal strip and the red sand strip that adjoins it. The former has thin, black soil that is generally of little use. The red sand can produce good sugar crops, but in times of drought (which frequently occur in Barbados) its productivity drops dramatically. The Bridgetown district has thin, relatively sandy soil and a long dry season. Although wholly unsuited for cultivation, its topography led to its choice as the principal town. East of the city is the St. George Valley. The soil here is a deep fertile black soil. The moderate rainfall makes it one of the most prosperous regions on the island, second only to the inland red soil district. The latter area has shallow soil and a severe lack of water.[3]

The climate the explorers encountered in Barbados was markedly different from that of their native land. Lying near the equator, it did not experience the seasonal variations that played so determinative a role in the shaping of English society. On this West Indian island, "winter and summer as touching cold and heat differed not." Temperatures ranged from 75 to 79 degrees Fahrenheit, with extremes between 90 and 65 degrees, depending on the time of year and where one happened to be on the island. The warmth of the Barbadian climate could easily support most tropical and semitropical vegetation, which led many to optimistic speculations about the colony's future.

Behind the beauty and fertility of Barbados's natural environment some unpleasantness lurked for man. Visiting the island four years after its settlement, Sir Henry Colt wrote of a multitude of land crabs. They were of a pale-yellowish color and had the bedeviling habit of pinching the colonists as they slept at night.[4] Thomas Verney told a no doubt exaggerated tale of how men full with drink would pass out on the island's roadways to be found the next day eaten alive by the crabs.[5] Colt also complained of an abundance of small gnats by the seashore. They appeared at dusk and began biting the settlers with such ferocity that no rest could be had "without a fire under your hammock."[6] Then there was the shortage of fresh water. Barbados has no rivers or lakes, and it is often beset by prolonged periods of drought.

Englishmen also found the general warmth of the Barbadian climate troublesome.[7] The English concept of man's physical nature in the seventeenth century was still strongly influenced by humoralism, the dominant theory in classical medicine. The history of humoralism dates back before the Hippocratic writers of ancient Greece but reached its fullest articulation in Galen, the Roman physician. According to humoralistic thought, changes in the weather could be pathogenic. Loss of health could occur, not only during the normal variations of the season (summer to autumn, autumn to winter), but also by abrupt downward or upward movements in temperature, unusually wet or dry periods, or migration from one climate zone to another. Since Englishmen were native to a cool climate, they came to blame virtually all their medical problems on the heat of the Barbadian climate, in spite of the fact that there were no "tropical" diseases (malaria and yellow fever, to name two) that were endemic to the island—and that the settlement ultimately proved to be as healthy as Boston or Philadelphia on the North American mainland.[8]

As a veteran traveler in the tropics, long before he set foot on Barbados Powell must have formed some thoughts on the wholesomeness of warm temperatures for Englishmen; it would be surprising if he did not agree with the popular sentiment that hot weather made men from cool climates ill. But Powell was so favorably impressed with all he saw that he immediately left the island and sailed back to England with plans to form a company to undertake its settlement. If he followed the tactics employed by others who sought to promote colonies in warm climates, he probably tried to dismiss fears in the mother country that the island was unhealthy by pointing to factors that kept the temperature at a level that was not pathogenic for Europeans—trade winds and the like.[9] He could further obscure the health issue by placing considerable stress on the natural barriers on the Atlantic coast that would help protect a fledgling colony from Spaniards and other invaders. In describing the island, he could speak well of its fertility, the absence of any landlord, and how it offered prospects of sizable profits if it were properly cultivated. In discussing profits, Powell would touch upon a subject that was of the greatest concern to potential settlers. It was the possibility of growing rich in Barbados that overcame anxieties about Spanish hostility, worries about the unwholesomeness of the weather, and a host of other fears and phobias that might have kept Englishmen at home.

After returning to England in 1626, Powell organized a company to finance his colonization scheme. His partners were the two powerful merchant brothers, Sir Peter and Sir William Courteen, their brother-in-law John Mounsay, and Powell's brother Henry. In 1627 they outfitted the ship *William and John* for an attempt to settle Barbados. The expedition reached the island on February 17, 1627, and eighty settlers disembarked to try their fortune in the New World.[10]

These pioneers and those that would later follow them into the West Indian frontier carried in their cultural baggage a comprehensive world view, whose norms and values were rooted in English culture. They were not, however, commited to any vision of reproducing the English world in the Caribbean. Their goal was largely material, and many Old World institutions and customs were lost in the search for wealth. Almost immediately a plantation society began to sprout on the island. By 1650, roughly two decades after the colony was founded, it was plain to any observer that although here and there this new community bore some resemblance to its English parentage, the offspring was a vastly different social system.

English society in the first half of the seventeenth century was poised somewhere between the medieval and modern worlds, if the two can be thought of as opposite poles. It was a time of major advances in science and technology and the birth of a world economic order. The colonists who came to Barbados were as much influenced by the new currents in English society as they were unsettled by the breakdown of traditional relationships.

In early modern England, the patriarchal family was the basic institution; the workshop, the farm, the landed estate were essentially family enterprises. To the seventeenth-century Englishman family meant more than husband, wife, and children. It also included all household servants. It was common to find the poorest of English families with at least one servant, but the household of the upper classes had the largest number of bondsmen.

The pervasiveness of servants within the English household was due to the custom of apprenticing children. In England apprenticeship was a means of social mobility. When a child reached the age of ten, lower- and middle-class parents usually contracted to have the child raised in another family. Under the guardianship of a surrogate father, the child matured into an adult, and he learned the occupation that his parents chose for him. From his master the servant also received food, clothing, lodging, and a good correction if his behavior warranted it. In exchange, it was expected that the apprentice would give his master loyal and obedient service. Apprenticeship lasted for approximately seven years. Once this period was over, the child, now a mature adult, was free to leave his master's household.[11]

Unless the apprentice was blessed with an inheritance, he was generally in no position to begin his own family, and so he became a journeyman. Journeymen were also considered to be servants. They worked from sunup to sundown in the household of a master tradesman, receiving wages and meals for their labors. Unlike apprentices, however, journeymen usually did not sleep in their master's house, and thus they had greater freedom. After the journeyman earned or inherited the means to settle himself as a master of his own household, he would ordinarily take a wife and begin a family. The task before him was to organize his household into a productive unit, see to the education of the young, and look after the moral character of all the members of his family.[12]

It is possible to divide English families into grand and petty patriarchs. Grand patriarchs (crown, nobility, and gentry) were the governing elite of England. The social and political bonds that existed between them and petty patriarchs in the seventeenth century were still very much a product of England's medieval past.

Landholding patterns, inheritance practices, and the structure of civil authority meant that in medieval England power drifted into the hands of the oldest male of landed families and that in every family, village, and manor there was a constant struggle to win the approval of, or establish some reciprocal claim upon, a grand patriarch who had land and offices at his disposal. Given the dependency of lesser men, individual loyalties to particular lords tended to hold greater place than obedience to central authority and devotion to the common good.[13] What developed between lord and subject was a relationship of good lordship—a reciprocal exchange of attendance, deference, respect, advice, and loyalty for protection, patronage, and a general interest in the well-being of subordinates.[14]

The allegiance that big landholders could command in English society did not mean that their powers were boundless. Beyond the sense of duty that emerged out of the notion of good lordship, there were definite advantages to having a retinue of faithful servants, and these benefits had a restraining influence on the English elite.[15] In this sensitive area, the guiding rule, underwritten by both the church and the crown, was that landlords should govern so that tenants would not be oppressed but so that they could live according to their "sorts and qualities." The church portrayed the lord/subject nexus as part of a hierarchical arrangement created by God that no mortal should attempt to alter. All the way up the social scale there rose a graduated ladder of dominance and subordination—children deferring to their parents, wives to their husbands, servants to their masters, and petty patriarchs to grand patriarchs. This great chain of being, it was argued, was the source of all the prosperity and internal stability that Englishmen enjoyed. Everyone was admonished to obey their superiors, even if the latter seemed tyrannical, and the elite were warned that they risked the wrath of God and the destruction of their family and country if they failed in their duties.

In spite of the dire predictions, and the general determination to preserve the status quo, by the reign of the first Stuart, James I, whose coronation took place in 1601, the great chain of being as it was constituted in medieval England had undergone some important modifications.

Over the centuries from the reign of William the Conqueror to James I, the centralizing tendency of the crown, particularly during the Tudor dynasty (1485–1603), did much to alter the structure of English societal relations. Gradually, English kings and queens took over many of the civil and judicial functions that were once performed by medieval lords. In theory, at least, the crown sought to institute public policies that would sustain a large population whose material interest would cause them to defend Great Britain from foreign aggression. In a sense, the medieval notion of good lordship was being applied on a national scale. Such diverse political commentators as John Hales, Thomas Wilson, James Harrington, Thomas Hobbes, and John Locke drew a connection between the ability of the state to encourage a materially contented population and the continued survival of an independent English nation.[16] The writings of Francis Bacon on maintaining high morale in the English army can be taken as exemplary of this broad consensus. "The number (itself) in armies import not much," he warned, where the people are "weak of courage." This failing of courage could happen if the state did not take heed that the upper classes did not "multiply too fast." Allowing too great a percentage of the society's wealth to fall into the hands of a minority would make the common subject "grow to be a peasant and a base swain, driven out of heart." Their debasement would mean that not a "hundred poll would be fit for helemet, especially as to the infantry, which is the nerve of any army and so there will be great population and little strength." Bacon advised the government to follow the example of Henry VII, who "bred a subject to live in convenient plenty and no servile condition."[17]

Following this logic, in early modern English political culture, yeomen were idealized as the most desirable element in society, the backbone of the army, and the mainstay of domestic tranquility. They were the elite of the petty patriarch class. For the most part, yeomen were prosperous tenants—often with several servants in their family. Some were rich enough to contemplate purchasing a coat of arms, and by that act join the lower ranks of the gentry. Their daughters, owing to the large dowries their fathers could afford, were frequently courted by impoverished men of higher status. In England there was broad agreement that the comfortable existence that yeomen enjoyed bred in them moderation and a commitment to the perpetuation of the social order. English political thinkers were always quick to note that a decisive element leading to English victories over the French in the Hundred Years' War was the presence of yeomen among their troops and the enlistment of disinterested mercenaries in their adversary's army.[18] Unlike yeomen, poor men were thought to be prone to crime and rebellion.[19]

In Barbados planters worked from the assumption that their slaves, given their legal status and the deprivation under which they lived, could not be relied upon to defend the plantation system in an emergency. Rather than improve the material circumstances of slaves, and thereby cultivate in them a genuine loyalty to the society, planters looked to England for the wherewithal to maintain the integrity of plantation society. This decision had far-reaching political implications.

In spite of the recognized need to sustain the population at an acceptable standard of living, on the eve of colonization English society was being unsettled by widespread economic dislocation. Although the difficulties never reached the disastrous proportions experienced by France, they were nevertheless of sufficient severity to stimulate profound changes in the nature of English life. Sensitive to the political and social implications of economic uncertainty, the English governing elite treated the instabilities as a serious emergency that required immediate attention. They were, however, deeply factionalized along religious and political lines, and the economic crisis only tended to widen the breach among them. It was in this period of general instability that the colonization movement began.

The elites of Barbados represented a new mentality that was arising out of the chaos that gripped Stuart England. It was individualistic, competitive, and highly materialistic.[20] Forswearing traditional notions of good lordship and medieval Christian values, it was quite prepared to allow material interest to serve as the yardstick for acceptable and unacceptable social behavior. The possibilities of amassing a large fortune brought men of this persuasion to Barbados.

In the first group that came over with Powell was Henry Winthrop, nephew of Emmanuel Downing and son of John Winthrop of Massachusetts fame. It is clear from his letters to his father and uncle that Henry intended to make his fortune in Barbados. He told them that he was presently planting tobacco that he hoped would be "very profitable." He, along with others, were promised by Captain Powell the attractive wage of a £100 a year, besides "their servants share they were to have the benefit of them." Winthrop planned to have two or three servants sent over to him each year, and "there always to have a plantation of servants."[21] Being a good Puritan, his father was not at all pleased with his son's activities. He admonished him for his "great undertakings, having no means." "Solomon sayth," he reminded Henry, "he who hasteth to be rich shall surely come to poverty." He would have been wiser to have contained himself in a moderate course for the three years. "By that time," the elder Winthrop wrote, "by your gettings and by my help you might have been able to have done somewhat." The "fruits of your vain overreaching mind," he warned, "will be your overthrow."[22] Henry may have been incautious in his enterprising, but he was typical of many who came to Barbados.

Samuel Winthrop, brother of Henry, proposed in 1648 to go to Barbados, where he thought he "could live better than in other places."[23] He conceived his "greatest strait" would be at his settling, his "stock being small." Thomas Verney, the profligate son of Sir Edmund, arrived in Barbados in 1637, at which time he bought 100 acres of land. When describing his circumstances to his father, he wrote that if he were sent the supplies that he requested there would be no question about his being able to raise a fortune on the island in a few years. If the provisions were not forthcoming, the younger Verney told his father, there would "be no staying for me in Barbados."[24] Verney's words are instructive; they reflect the purpose for coming and the reason for remaining in the colony.

Most seventeenth-century travelers who stopped at Barbados were immediately struck by the unusual acquisitiveness of the people, a trait that seemed to know no moral boundary. Governor Willoughby wrote to his wife in 1652 that the woman bearer of his letter was "one who challenged acquaintance of me upon your score, which caused me to give pass for her sugar, custom free." The governor had no recollection of the woman; nor would he have believed her had she not named one of his children. Disturbed by the possibility of deception, he told his wife that if the letter reached her safely he would not "be much troubled at the cheat; for it is frequent here to have tricks put upon one of such kind."[25] In these few words Willoughby called attention to the unbridled acquisitiveness that characterized Barbadian society. Many others besides Willoughby commented on the unchecked materialism that pervaded the colony. Thomas Walduck thought that Barbadians had "no moral honesty." In an acrostic on the island's inhabitants, he opined that "for one honest man" on the island there were "ten thousand knaves."[26] Father Antoine Biet, visiting the colony in 1654, was one who came remarkably close to exposing the source of this materialism. "In speaking of morals," he wrote, "extravagance is very great among the English in these parts. They came here in order to become wealthy."[27] It was no accident that Biet mentioned the islanders' efforts to enrich themselves while trying to shed light on their morals. The obtaining of profit was, as Morgan Godwin contended, "the most operative and universally owned principle" of this society.[28] Into this "wilderness of materialism," as one historian has described it, Africans were forcibly dragged to serve the ambitions of a few large planters.

Prior to the fifteenth century, the African communities from which the founders of the Afro-Barbadian community were drawn were very much isolated from Europe, and no African could have predicted that some of his neighbors would be living with Europeans in lands across the ocean. There were a few trade routes with the European world; most connected the Gold Coast, with its rich preserve of precious metals. But the topography of sub-Saharan Africa (in the north, the Sahara Desert, just below it a savanna region infested with the tsetse fly that preyed on draft animals, and then a dense tropical rain forest) acted as a natural barricade limiting contact between Africans and Europeans.[29]

Recapturing traditional African societies as they existed in the seventeenth century cannot be done with the same detail that can be done when writing about the turmoil that beset early Stuart England. African communities were preliterate societies; their histories were passed from one generation to another orally. Those events that became part of a community's oral tradition were often an occurrence of large societal importance, and it is difficult to sew together the fabric of everyday life from such threads.[30] Furthermore, not all Africans in Barbados came from a single country—or region, for that matter. Slaves were taken from West and Western Central Africa—an immense area with wide ethnic diversity. There were Ibo, Ijo, Itsekiri, and Ibibio from the Niger Delta; Ashanti, Fanti, Ga Akwamu, Akim, Agona, Obutu, Mamprusi, Dagomba, Mankansi, Talense, Isalla, Lober, Ardas, Popos, Whydanhs, and Yorubas from the Slave and Gold Coast; and Kongo, Kuba, and Ngondo from the Congo and Angola.

This diversity does not rule out the possibility that there were values and unconscious, cognitive orientations that these peoples may have held in common. It is quite possible that the various African cosmologies shared some basic assumptions about social relations and the way the world functioned phenomenologically. The common orientation to reality may have tended to direct the attention of Africans to similar kinds of events, even though the way they responded to these events may have been quite diverse from one society to another.[31]

While sub-Saharan Africans probably held in common certain cultural understandings and associations—and these similarities were to help shape plantation culture—such commonalities in orientation were not enough to bridge the heterogeneity that existed among these imported workers. Africans came from social systems that inculcated exclusive, ethnic, and chauvinist ideas of self. In traditional Africa there were hundreds of ethnic groups, each deeply committed to its own preservation and existence, and ethnic conflict was a fact of life. In the New World, ethnic identification and, subsumed in that, the cultural and linguistic differences that divided African peoples, inhibited a collective (pan-African) response to their enslavement.[32] Indeed, in Barbados the diversity of Africans and their particularistic sense of self facilitated their incorporation into the island's plantation system and eventually permitted important changes to occur in their cognitive orientations. The Afro-Barbadian community, as a coordinated cultural and social entity, is a New World phenomenon, though fragments of the African past survived in some significant areas.

Speaking broadly, virtually all the Africans who were forcibly transported to Barbados came from societies that organized themselves in terms of kinship and lineage.[33] Some were composed of a single descent group in which everyone in the community traced their origins back to the first man created by God or to a founding ancestor of extraordinary, if not supernatural, abilities. This group could usually lay claim to a specific geographic region as its country or land. In oral traditions and myths, a common history was articulated that provided the community of relatives with an ethnic identity that established its distinctiveness and superiority to outsiders.

There were also African societies that were multilineal in character. The latter case was frequently a product of conquest (one descent group subjugating another) or the outgrowth of peaceful negotiations. In either instance, multilineal societies would usually synthesize traditions, rituals, and institutions that explained and legitimated the union of the various descent groups. The intention was to transcend kinship identification and to create concerns for the larger community. The degree to which multilineal societies were able to foster loyalties to the larger group varied from one community to another, but rarely was allegiance to one's lineage totally destroyed.

It would require more space than is available here to fully describe the intricacies of African kinship systems as they related to social, economic, and political behavior. Suffice it to say that in Africa kinship was determined by blood and betrothal (engagement and marriage). The kinship system was a network that reached out horizontally in every direction. In some manner it embraced all the people of a given community. In Africa kinship terms such as "father," "mother," "grandmother," "brother," and "sister" were not used in a strict biological sense. In some groups, for example, an individual would refer to, and treat his mother's sister (aunt, in Western kinship classification systems), as his "mother"; or, his father's brother (uncle) as his "father." Consequently, it was quite possible for a person to have hundreds of "fathers," "mothers," and "uncles." With some qualifications, each person in a community was a "brother" or "sister," "father" or "mother" (or some other such relation) to everyone else in the society, and there were many terms to acknowledge the exact nature of the kinship bond between individuals.

Kinship in Africa was also reckoned vertically to include the dead and those yet to be born. In African cosmologies, death did not mean a termination of existence but rather a transportation into a spiritual dimension inhabited by other community ancestors. In this spiritual world, the "dead" could still have their influence on the physical plane.

It was thought that they could mediate with the gods to secure prosperity and health for the living community; as a result, ancestor worship was a common feature in African religions.

In many African societies, children were taught the genealogy of their descent group (both vertically and horizontally). The genealogy gave them a sense of lineage, of connectedness and historical belongingness, and a sacred obligation to extend their genealogical line. Within the community, lineage, particularly when the society was composed of more than one ethnic group, helped to inform political decisions. It was also a factor that worked to perpetuate a rigid and formal code of conduct that regulated daily intercourse between individuals. When strangers of the same tribe met, for instance, one of the first things they attempted to sort out was how they were related to each other. Once they understood how the kinship system applied to them, their behavior followed the patterns set down by the society. "Brothers," adhering to group customs, might treat each other as equals, or "younger brothers" would defer to "older brothers"; a "nephew" might be expected to show deference to an "uncle." It was possible that from that moment on they would refer to each other by the kinship terms of "brother," "nephew," "uncle," or "mother," with or without using their proper names.[34]

Any attempt to sketch out the elaborate etiquette that surrounded kinship relationships in Africa must take some notice of African religious beliefs and activities.[35] Although fundamental concepts such as the continuation of human life after death, the existence of the spiritual world, and magic and witchcraft were common to African religious systems, each society had its own distinct religious rituals, symbols, and dogma. There was no universal church to which all Africans belonged; nor did communities try to proselytize outsiders. Ordinarily, a person could not be converted from one religion to another. He had to be born into a particular society in order to participate in the entire religious life of the community.

Common to African religions was the idea that sickness and misfortune (including natural disasters) were a consequence of the malevolent use of magic. The ability of one tribe to conquer and subjugate its neighbors was most often explained in terms of the victor's superior use of witchcraft and magic. Religious movements whose purpose was to overcome the evil doings of witches and sorcerers frequently evolved in traditional African communities. These movements began with a charismatic leader, who experienced visions for several years before he or she convinced the local community that they were true. When the community accepted the message, it observed the newly proposed rituals that were expected to bring fortune to the village and do away with affliction and witchcraft. A charm constituted the centerpiece of all new movements, and its purpose was to protect the community against disease, death, and misfortune. Some charms, for example, were thought to make an individual invulnerable against the sorcery of a foe.

Convincing the community of the efficacy of the movement took time. One did not see a single-day victory, but rather a period of ebullience with new cult members dancing and singing at night or during part of the day to attract others. Once the first congregation was formed (whether it comprised all members of a village or not depended on the local social structure), it developed an internal organization. Socialized religious duties were allotted to "dignitaries." These officials were, in a Weberian sense, the charismatic lieutenants of the original leader.

The movement spread as neighboring settlements became convinced that it offered better protection against misfortune. At this juncture, people rushed to the nearest village where the movement had been adopted and asked to be initiated. The initiation presupposed a dramatic theaterlike production. Old charms were discarded; the new movement was brought in with much dancing and singing. People were purified in one way or another (rites of communion, confession and incorporation), and the prospective new local leaders learned their "specialties" from the initiators. The whole process lasted at least several days. The new movement brought harmony and peace, and therefore it eliminated witches at the same time.[36]

Misfortune, sickness, and death were also thought to be caused by departure from traditional modes of behavior, such as violations of established kinship relations (adultery, incest, or disrespect to ancestors). Here it ought to be understood that every society had its own ideas about what constituted unacceptable behavior. In some communities a man might be considered adulterous or incestuous if he ate food prepared by his "mother-in-law"; others regarded the birth of twins as a sign of a wife's infidelity to her husband, and the woman and her children might be put to death. In cases where improper social conduct occurred, it was generally believed that the whole group would suffer as a result of the malefactors' behavior, so that the community at large would seek to punish the offending individuals. Thus the etiquette of kinship relationship was supported by deeply held religious convictions, and transgressions of proper conduct could lead to severe punishment.

African societies were very formal ones; but this fact should not be taken to mean that Africans were moral or legal automatons, blindly following traditional principles of ethics. As in all human communities, men and women exploited and stretched societal rules for personal or joint advantage. In African societies, as powerful an agent as loyalty to one's descent group was, there are many examples of its succumbing to market forces in the era of the Atlantic slave trade. It is perhaps best to view the restrictions imposed by religious convictions and codes of kinship relations as the moral values that communities used to judge socially acceptable behavior.

Kinship was a force that affected the economic life of African communities.[37] The most basic social and economic unit in Africa was the household. It generally consisted of three or four generations from grandparents to grandchildren, and perhaps to great-grandchildren. The household was under the domestic rule of a single man (or, occasionally, a woman). This person might also be the individual who represented the nuclear group in political councils or political-religious ceremonies that included several families. At least in principle, the head of the household organized the family's labors, skills, and reproductive capacity into a self-supporting unit of producers and consumers. In practice, however, the household was able to exist only within a community of families who helped each other in economic and defensive ways.

As in most preindustrial societies, the number of persons in a household set some limits on its productive ability. In an attempt to increase the size of the household's labor force, Africans frequently procured slaves. By and large, those who properly might be called slaves in African societies were outsiders (i.e., those without kinship ties in the community). They were acquired in several ways. Some were children who were considered to be supernaturally dangerous by their community, and so they were abandoned by their parents. Criminals, those who broke societal rules, might also be sentenced to bondage and sold to another group. The acquisition of bondsmen as a consequence of interlineal wars and raids was widespread, but it was not until the coming of Europeans that it became the most significant means of procuring slaves in West and Western Central Africa. Yet bondsmen were a valuable commodity in African societies even before the arrival of Western slave traders. Since they had no kinship ties in their master's society, and thus were outside the etiquette that regulated the interaction of people within African communities, they had to look to their owners for protection, and so they tended to give loyal service. For masters, the slaves' versatility, mobility, and capacity to support and reproduce themselves made them important assets. They provided extra wives and children to expand a kin group, laborers to till the fields, retainers in the compounds, soldiers, trading agents, and personal servants. In some communities, slaves developed into a powerful class in their own right. Male slaves in particular had a wide range of occupations open to them, and they were often placed in positions of authority over free people. Purchased as a child, a slave could become a member of an elite class of quasi-military administrators, such as Wolof Tyeddo, who were similar to the Turkish janissaries or Eygptian mamelukes. In other societies, however, bondsmen could be subjected to conditions that challenged life itself. Moreover, however important a slave's status might reach in a particular society, he was still under significant disabilities as an outsider. To become fully incorporated into the society, he had to establish kinship relations in the community; in so doing, he became subject to, and protected by, the moral obligation that affective relations imposed upon individuals.[38]

These, then, were the main attributes of African societies. A person had to be born into a community, and he could not easily change his societal membership. Each society had its own peculiar religious, kinship, and political systems. Together they created a particularist world view in which everyone held loyalty to their descent group specifically and to the community more generally as the primary social value. In this world few felt much responsibility for persons who were not considered to be relatives or who did not belong to the community. The lack of concern for individuals defined as outsiders meant that few restraints governed relations with aliens; it was quite all right to reduce the latter to commodities of exchange—their market value being of greater importance than their instrinsic worth as human beings.

The slaves that were brought to Barbados had no sense of themselves as a black or African people. In their minds, they were Kuba, Ashanti, Akim, or one of the other African communal groups represented on the island. Cultural, social, and linguistic differences separated them from the other men and women whose fate they came to share in the colony. For the African in Barbados, an ocean separated him from those who helped to reinforce his traditional sense of self. Tribal origins were not immediately forgotten, but for the slave, and more particularly for his children, the break with traditional African ways began aboard slaving vessels destined for the New World.

Before their transplantation to the West Indies, Africans and Englishmen lived in societies that, when compared with each other, reveal some general similarities and some notable differences. While it differed in structure and function, the basic social institution in both England and Africa was the household. In both regions, it operated in some manner as a workplace, a home, a hospital, and a school. Servants and bondsmen were commonly found in both the traditional African and English households, though it must be pointed out that Africans and Englishmen held vastly different views about who was or was not a bondsman and what social limitations this status placed on an individual.

In many African societies, the head of the household represented the members of his family at community political gatherings. Decisions reached at these events were deeply influenced by ideas of kinship, lineage, community, and religion. In multilineal societies efforts were made to create social relationships that embraced all members of the society. In England, at least in theory, the political voice of the members of a household was invested in its patriarch. There was, however, a strong sense that men of small means ought not to be trusted with political responsibilities; thus, there was a tendency to exclude them from positions of authority and to limit their participation in political affairs. Power was concentrated in the hands of the eldest male of landed families. It was widely felt that the country ought to be governed so as to maintain a yeoman class whose material stake in the society would make members of that class want to preserve the status quo.

Africans and Englishmen were devoutly religious peoples who believed that spiritual beings could influence events on the physical plane. Both accepted the possibility that certain individuals could perform feats of magic and witchcraft that could have good or ill consequences, though in England religion and magic were slowly giving ground to a rational material sense that facilitated participation in a world market system.

West and Western Central African communities evolved without much contact with Europe and Asia. This isolation meant that much of the science and technology that Englishmen absorbed from other civilizations were unknown to Africans. The latter also did not participate in the centuries of scholarly dialogue that were to produce the scientific advances of the seventeenth century.

Finally, Africans were not a homogeneous people, and though Englishmen in the seventeenth century were only beginning to articulate a nationalistic identity, they were certainly much more of a cohesive, sociohistorical group than were the people who were to become their slaves. In Africa and Europe race was irrelevant as a social variable—everyone was either white or black. In these Old World communities forces such as kinship and lineage on the one side and property and family on the other were the devices that separated social groups. It was in Barbados's slave plantation system that race acquired social meaning.

CHAPTER 2

The Birth of the Plantocracy

In the seventeenth century Barbadian society was organized around semiautonomous plantations. Before the 1640s it was not totally uncommon to find a plantation of 2,000 or 3,000 acres, but by the 1680s no one owned more than 1,000 acres—the biggest planters had at least 100.[1] Within the boundaries of these estates lived planter-patriarchs who owned and autocratically ruled over the servants and slaves residing on their property. Although the island is only 166 square miles, its dense tropical vegetation, even in the more cultivated areas, and its difficult terrain acted as natural barriers separating one plantation from its neighbors, creating a sense of physical isolation for plantation inhabitants.

The central institution on any plantation was the household. For those Englishmen who pioneered isolated farms on Barbados in the 1620s and 1630s, where masters and bondsmen frequently slept under the same roof, the English custom of seeing domestic servants as part of the family survived. As the number of servants increased on plantations and as individual planters gained the means to provide separate quarters for their bondsmen, the practice was for masters and servants to live in different buildings. Planters, however, continued in their habit of referring to all bondsmen as members of their *family*, and so in Barbados the words "household" and "family" were stretched to include servants who were not actually sleeping in the planter's house.

Slaves were also considered to be members of the planter's household. Henry Drax, for instance, spoke of "all the members of my *family* [italics mine], black and white."[2] The roots of this tradition were established in the late 1620s. Although in these early hours of settlement there were qualitative differences in the life-styles of servants and slaves, as the exclusion of the latter from the Christian church would seem to suggest, the small number of slaves in a Barbadian household in this era meant that in certain basic respects (amount of food and clothing received from their owners, working hours, and leisure time) the daily existence of slaves resembled that of the average Christian servant. Equally important, the colony's tiny African minority posed no threat to the social order, and planters had no reason to fear it. In this environment there was no impediment to prevent English farmers from extending the word "family" so as to include black slaves. Indeed, the planter's inclusion of all dependent labor, both black and white, into his "family" fit well with the fundamentally patriarchal society that emerged on the island. The plantation owner was the father and lord of his "household," and all servants and slaves looked to him for protection.

If properly qualified, the neologizing of the terms "family" and "household" in seventeenth-century Barbados can be of service to modern historians, as they can help to frame and label an important group of political, social, and economic relationships that evolved on the island. From the founding of the colony, a sizable proportion of its inhabitants were held by others as servants and slaves. By 1650 well over 85 percent of all Barbadians were dependent laborers, living in the households of approximately 2,000 planters. Because the central government in Barbados generally did not interfere in the management of plantations, particularly in those areas related to the governance and treatment of bondsmen, planters' ability to regulate the lives of servants and slaves was quite extensive. Their households functioned in some fashion as a hospital, a local government, a school, a workplace, and a home. Consequently, little can be understood about the character of political, social, and economic relationships in the colony unless they are analyzed against the backdrop of household organization on the island.

While the neologizing of the words "family" and "household" in Barbados so as to encompass servants and slaves has advantages for historical analysis, the historian would soon encounter difficulties if he completely followed the Barbadian example. The term "family" as it frequently came to be employed in Barbados had nothing to do with the human productive unit. It also obscures the fact that in a single "household" there might be many separate nuclear groups (husband, wife, and children). To avoid confusion, "family" is used here in discussing nuclear or biological kinship groups. Household members are the collection of people that planters had some legal control over, to some greater (servants and slaves) or lesser (wives and children) degree.

Speaking in broad terms, it can be said that the Barbadian political community was dominated by the heads of the largest households on the island. Even in the most liberal of times a person had to be a white male of at least sixteen years of age possessing ten acres of freehold in order to vote. Few on the island could meet all these qualifications. Indeed, from the first days of settlement, the heads of the biggest households formed a small plantocracy. Collectively, they owned most of the land and labor on Barbados, and they used their proprietorship to establish themselves as the colony's political elite.

Initially the composition of the plantocracy was unstable; men rose and fell from power and relative wealth in the wink of an eye. By the 1660s the plantocracy had solidified into an identifiable group of families. A significant core of those who settled plantations in the 1620s and 1630s did manage to escape the pitfall that consumed others, establishing their families as part of the colony's governing class, but when listed alongside those with whom they shared power in the formative years, they appear as a tiny minority.[3] In spite of this fluidity, what remained constant was the dominance of the plantocracy as a sociopolitical entity. Though poor whites agitated, and servants and slaves occasionally revolted against their masters, no segment of the island's population ever effectively challenged its authority.

The beginnings of big planter cohesiveness are to be found in the common material conditions that they shared in Barbados. In the 1620s and 1630s, land speculation, tobacco cultivation, lumbering fustic wood, and a general willingness to try anything that might turn a profit tended to characterize their economic life. They all owned large tracts of land, ranging from a hundred to several thousand acres. Unlike the big houses and well-cultivated lands their children were to possess, their plantations were desolate places, little more than small cutouts in the midst of a tropical rain forest. Working the fields of their estates was nearly half of the island's population, whom they kept as servants and slaves. The political bonds that held the plantocracy together were forged in their efforts to maintain control over the colony's servile inhabitants and in their endeavors to resist the centralizing tendencies of powers in the mother country. The principle that guided their actions was best summarized in the advice that Thomas Povey gave to his brother William, who lived in Barbados, which was to "farly and soberly consult your own interest."[4] Indeed, it is impossible to write about the beginnings of the Barbadian plantocracy without some sense of materialism as a historical force. The ships that transported English settlers to the island were filled with "dreams of prosperity," and the plantocracy and the society that it governed were deeply shaped by ambition.

It was from among those who could not find a secure niche in English society that the majority of voluntary migrants to Barbados were drawn. Many were from towns and wood-pasture villages, where medieval ties of subordination hardly existed or were rapidly waning by the time Barbados was being colonized. More likely than not, those who came from urban centers had only recently settled in towns. These were the masterless men who left their home parish in hopes of escaping hard times. They wandered into English cities, where they found virtually no opportunity for social mobility.[5] A disproportionate number of them were young males. Of the seventy-four persons who came over with Captain Powell, there were no women.[6] According to a register of passengers who left London for Barbados in 1635, only 6 percent were women; 91 percent were young men between the ages of ten and twenty-nine. No children under the age of ten and virtually no married couples were to be found.[7] More than a fourth of all Englishmen who came to Barbados before 1640 were unable to pay their own passage to the colony. Accordingly, they signed contracts that placed them in temporary servitude in return for passage to the New World. Their expectations were that, after a short period of service, they would be in a position to build a comfortable existence for themselves. Given the practice of apprenticeship in England, and the opportunities it afforded for advancement, their aspirations were reasonable. What they could not anticipate was that servitude in a Barbadian household would be much more rigorous than apprenticeship in an English household and that it would rarely lead to any real improvement in their material circumstances.

Not every Englishman who came to Barbados as a servant did so willingly. Unable to cope with a growing class of poor in their midst, in numerous instances town fathers rounded up vagrants and apprenticed them to the colonies. In 1636, for example, fifty vagrants were taken up in London, and bound by the lord mayor and aldermen as apprentices to serve in Barbados and Virginia.[8] Planters were never overscrupulous about how their servants came to them. They took convicts from English prisons and accepted without question victims of "spiriting" (kidnapping). Indeed, so common was the practice of spiriting that the verb to be "Barbados" was coined to describe the experience.[9]

Whether they came as servants or as men of modest means, there were few rags-to-riches stories to be told in the colony. Those who emerged as members of the Barbadian plantocracy had strong ties to the middle and upper strata of English society well before they settled in the colony. Some were younger sons of English gentry (Henry Winthrop and Thomas Verney) who were forced by inheritance customs to seek their fortunes abroad; others can be described as lesser gentry (James Drax, Reynold Allyne, William Hilliard, Thomas Gibbs); and there were close relatives of noblemen in the mother country (James and Peter Hay were related to the lord proprietors of the island, the earls of Carlisle; William Tufton [briefly governor of Barbados, 1628] was the brother of the earl of Thanet).

The connections that these men enjoyed in England was a key ingredient that contributed to their rise in Barbados. The patronage and influence of grand patriarchs in England stretched across the Atlantic and provided them with the credit to purchase land and provision by which to start a plantation. In 1637, shortly after his arrival in Barbados, Thomas Verney wrote home to his father requesting financial support to help him build a plantation on the island.[10] Henry Winthrop addressed a similar letter to his father.[11] After contracting for half of a 100-acre plantation, James Dering wrote his cousin Edward Dering, a man of wealth in the mother country, for help. James told Sir Edward that he was moved to get involved in the venture "out of assurances I had of your Worships love at all times shown unto me and the ready and willing help which I always received from your Worship." Echoing the optimistic predictions of Henry Winthrop and Thomas Verney, James calculated that he would be able to triple or quadruple his original investment. Without his cousin's help, however, he feared that he would be "utterly undone and should lose all that ever I had here."[12] Connections and access to credit were by no means a guarantee of success in the colony, as the careers of Dering, Winthrop, and Verney make plain; but, for the general masses who came out as servants or perhaps with sufficient resources to purchase a small farm, the lack of these was a prescription for disaster.

The Barbadian plantocracy was born close to wealth and power in England and moved in genteel circles, but few could boast of large personal fortunes before coming to the West Indies. They brought with them the same preoccupations with acquiring, preserving, and increasing land estates that were visible among the grand patriarchs of English society, and their concerns helped to shape the landholding patterns that would emerge as one of the distinguishing characteristics of the island's plantation society.[13]

The plantation as a socioeconomic institution dominated Barbadian history from the outset of colonization. This fact is all the more remarkable when one considers that structurally the plantation had no equivalent in English or African society. The residency of labor on estates, the authority of patrician landlords, and the imposing presence of the big house and the manor house may suggest a resemblance between the Barbadian plantation and the English manor, but the similarities are superficial. The plantation had no villeins, no manorial courts, and was not cultivated by open-field agriculture. Of great importance, none of the workers of the plantation had legal or customary rights to estate lands as occurred in the manorial system, and the idea of reciprocity that developed between medieval lords and their tenant/subjects completely lost its meaning in Barbados, as planters established their authority primarily on their ability to coerce the colony's working population. If the plantation was not manorial in character, it certainly did not resemble the modern forms of agriculture that were taking hold in England. It is true that it was organized to produce for a world market, as were modern farms in the metropolis, but the plantation's use of servile labor had far-reaching implications that served to differentiate it from modern farms in England where free labor was employed. Most important of all, the plantation was in part an admixture of African and European elements—a distinct social system created by conditions and market forces that were present in Barbados.

The members of Powell's Company planted the first seeds that were to sprout into a plantation community when they carved out five large farms on the island.[14] They attracted workers to the colony by promising high wages—Henry Winthrop was offered £100 a year. As an added inducement, they promised that if any of their employees brought servants to the island they were to have whatever wages their servants were entitled to.[15] This arrangement did not last long. Those employees who could afford to bring servants to the island quickly realized that they might earn more if they could cultivate the land themselves; consequently, the five original plantations gave way to twelve.[16]

Unfortunately, after this first division there is no reliable list of landholders in Barbados until the census of 1684. In *Some Memoirs of the Settlement of Barbados*, written in 1741, William Duke left a list of 758 planters who owned land in Barbados in 1638, which purportedly was based on documents that are no longer available to us.[17] When this register is compared against a list of persons who purchased land in 1636, approximately 50 percent of those who bought land in 1636 failed to make Duke's roster.[18] While Barbadian society was still extremely fluid in the late 1630s—and it is conceivable that a number of those who purchased land in 1636 had left the island by 1638—50 percent seems high. The sense is that if Duke's list had been accurate it would have been two or three times as long, putting the total number of proprietors at around 1,500 to 2,000.[19] This would allow for some smallholders but would recognize that big planters (roughly 10 percent of all landholders) owned more than 50 percent of Barbadian land.

While the sources permit only a hazy judgment to be made about the extent of individual holdings, it is quite plain that from the beginning of colonization there was widespread speculation, with a general trend for land to be monopolized by a few big planters. Philip Woodhouse owned at least 100 acres as early as 1630 and was still purchasing more in 1636. Francis Whitefield had 120 acres, and George Buckley's Parris Spring Plantation consisted of 300 acres. The Three Spring Plantation owned by Francis Skeete was huge, more than 4,500 acres. Edmund Read, a powerful political force in the colony from the mid-1630s until the 1650s, held at least 500 acres in the 1630s. Capt. Anthony Marbury, a member of several early councils, had at least 120 acres. William Hilliard possessed 610 acres and bought 430 additional acres in 1639. The Hawley brothers, Henry (governor, 1630–38) and William (deputy governor), owned over 1,000 acres. All these plantations were dwarfed by the 10,000 acres granted to five London merchants.[20] It seems clear that the monopolization of Barbadian land took place in the first decades of settlement and that the total acreage in the hands of small planters was always well below that controlled by large estate owners.[21]

For every 10 acres that a planter possessed, he was required by the lord proprietor (after 1634) to maintain a servant to cultivate the grounds; otherwise, he would forfeit his property for nonmanagement. This measure was supposed to ensure that there would be recruits for the island's militia. It also encouraged large estate owners to lease a portion of their holdings to freemen, and thus a sizable subtenant population emerged in Barbados. The leases were for a specific period of time; the rent was fixed (sometimes paid in specie, at other times in kind, usually in cotton or tobacco); and the renewal of a contract was not guaranteed. The size of a tenancy varied greatly. It was unusual, however, for them to be less than 1 acre or greater than 50. Tenants ran their own households and cultivated their own plots. Those who could afford to do so maintained a few servants and slaves.[22]

In the first three decades of settlement there was an acute shortage of labor in Barbados. As one planter put it, "a plantation was not worth much unless it had a good store of hands to work it."[23] If we could erase the deterministic features of their hypothesis, Barbados in the 1620s and 1630s would be a classic example of the land/labor imbalance so extensively written on by Wakefield, Nieboer, and more recently Domar.[24] Their thesis accurately describes the island's labor market at the time of colonization. Unlike the Spanish mainland, Barbados had no native population that could be intimidated into working for the benefit of the settlers. If profitable agriculture were to develop on the island, a labor force would have to be imported from abroad. Nor could people be simply carried to the colony and permitted to barter their labor as they pleased in a free market. Given the possibility for geographical mobility, there was no reason why a planter denied land in one colony could not migrate to another where he could gain a freehold or some form of lucrative employment. With each man able to possess sufficient land to satisfy his basic needs, there was no incentive for the colonists to sell their labor except for very attractive wages. But reliance on free labor, with the high wages it could command, could have proved so costly as to inhibit capital formation. Barbadian planters resolved this land/labor problem, as did some of their neighbors in other colonies, by restricting the geographical mobility of labor through some form of servitude.

Unfree laborers were the majority in the island's work force from the founding of the colony. Planters took on bondsmen without regard for race, creed, or national origins, though they were by no means "equal opportunity employers." They discriminated between Africans, whom they treated as perpetual slaves, and Europeans, who were enlisted as indentured servants.

For all the importance that slavery was to have in the development of Barbados, and by example in the rest of the English Americas, its beginnings in Barbados was marked by little fanfare. During the original voyage of settlement, Henry Powell captured ten blacks from a Portuguese prize. Aware that a shortage of labor was going to be a crucial problem facing the infant colony, Powell took his windfall with him to the island.[25]

The number of blacks in the colony seems to have increased rapidly. In October 1627, a mere eight months after Powell reached Barbados, Henry Winthrop wrote home that of the 100 persons that now inhabited the fledging colony 40 were slaves. Winthrop's figures are open to some question, but he leaves the clear impression that blacks were a significant part of the island's population.[26]

In the 1620s Negroes may have represented a sizable proportion of the colony's inhabitants, but it was not long before indentured servants surpassed them in number. Considering the early use of Africans, it would not appear that an aversion to blacks or slavery was responsible for the shift in the labor force. Indeed, if anything, the evidence seems to suggest that planters preferred black slaves to Christian servants. Slaves, after all, could be held in perpetual servitude, with their children following the status of their parents, assuring the colonists a permanent class of laborers. Besides, following the dictums of humoral theory, English settlers firmly believed that blacks were biologically better suited to work in the heat of the Barbadian climate than Europeans, and hence were more productive and cheaper to employ than white workers.[27]

The principal reason for the use of servants over slaves in the late 1620s and 1630s was the high cost and scarcity of the latter and the relative plenitude and cheapness of the former. Between 1627 and 1638 Barbados stood at the edge of economic collapse; the colonists could barely muster sufficient trade to keep the settlement stocked with essential commodities from home. If English merchants made infrequent stops at the struggling colony, slave traders had absolutely no reason to visit the island. The few blacks who reached Barbados in this period came as private prizes or were brought to the island by their masters who had migrated from failing English colonies in other parts of the Americas. Planters had to rely on a few supply ships from the mother country to bring them servants to labor on their plantations.[28] As one might suspect, the number of Christian bondsmen brought over in the 1630s was not enough to satisfy the comparatively small demands for workers that existed during the years of economic stagnation, but they were certainly more plentiful than slaves.

On the eve of the sugar revolution, 1638, there were roughly 2,000 servants and 200 slaves on Barbados (the total population was about 6,000). The majority of these bondsmen were in the possession of big planters. The few plantation inventories that have survived from the 1630s suggest that a planter who owned 100 acres of land usually held

5 to 10 servants and 1 or 2 slaves.[29] A tiny minority had substantially more bondsmen. William Hilliard came to Barbados with 60 servants.[30] The Hawley brothers had over 90 bondsmen.[31] What kept small planters from acquiring servants and slaves was the cost of buying them. In the late 1630s, a slave was valued at £25, and a servant who owed four or five years of service was priced at £12.[32] Tenants and small planters were rarely able to pay such a sizable sum for a laborer.

With big planters able to call a third of the Barbadian population their chattels, and with another third of the island's residents their tenants or small planters, it is clear that the average Barbadian was in no position to question the authority of the plantocracy.

In Barbados, servants entered a world in which little restraint was placed upon patriarchs in their dealings with bondsmen. Cut off from the prying eyes of neighbors, plantations may be seen as tiny, isolated hamlets where heads of households governed their servants without much external interference. The major extension of patriarchal authority occurred during the period of extremely weak central government under the Powell Company.

Shortly after the division of the original five plantations into twelve, the planters elected John Powell, the younger, governor of the settlement. Governor Powell's authority over individual plantation owners was minimal. Curiously, despite the new receptivity of the crown to efforts to begin colonies in the West Indies, the Powell Company began its enterprise without a patent or commission from the king. In short, all those that came out with the original party and later joined it were squatters. It is entirely possible that these adventurers had no expectation of planting a permanent settlement. Following the example of numerous Englishmen in the West Indies before them, they might well have intended to erect a few temporary plantations in hopes of making a quick profit. More likely, the company may have decided to begin settling the island without royal approval, as its occupation might strengthen the earl of Pembroke's claim to the island over that of his chief rival, the earl of Carlisle. If that was its plan, members of the company failed to promote Pembroke's cause, and the absence of a patent or commission left the syndicate without any claims to govern the independent plantation owners.[33]

Given his precarious legal position, Governor Powell could not, even if he so desired, make and enforce laws that would have protected servants from their masters' avarice. The impotency of the government, coupled with the absence of community restraints, gave planters a free hand over the bondsmen in their households. By the time strong central government emerged in the colony, a tradition had already taken root in which plantation owners ruled their servants with little outside regulation.

Recognizing that he would not be able to finance the settlement and protect these far-flung territories, about five months after the Powellites settled Barbados (July 2, 1627), Charles I granted James Hay, first earl of Carlisle, letters patent making him lord proprietor of the "Caribbee Islands," named after the Carib Indians who inhabited those islands situated between ten and twenty degrees north latitude, which included Barbados.[34] In full reversal of the long trend of reducing the power and authority of the aristocracy in England, Charles I reached back into England's medieval past and, for all intents and purposes, made James Hay a feudal lord. Carlisle was to hold these Caribbean lands of the king by knight service with as ample royalties and jurisdiction as ever possessed by the bishop of Durham within his bishopric or county palatine. By the terms of the patent, Carlisle could make laws, select clergymen, erect courts, appoint judges, enforce obedience by corporal punishment or sentence of death, and generally exercise the authority of captain general for the crown. As lord proprietor, the earl could dispense land on the island as he liked and create titles; he also had the right to charge an impost on all imports and exports.

Since the days of William the Conqueror much had changed in England, and the powers granted to Carlisle, whatever the expectations of the king, were not likely to result in the feudal relationships that once existed in the mother country. Land tenure by knight service, for example, no longer meant that a tenant of the crown had to supply knights for the Royal Army, and though Carlisle was charged with providing for the defense of the Caribbean colonies, there was nothing as visible as ancient knight service to demonstrate his compliance. Furthermore, the ties that existed between landlords and tenants in medieval England had given way to a purely economic relation; and the earl, at least in his West Indian possessions, adopted a modern attitude toward his tenants. Still, Carlisle had the power to build a strong personal following in the Islands of Carlisle Province, as he called his Caribbean holdings. Through his right to grant titles, offices, and land, coupled with his control over the church and the state, he could have encouraged a society of small farmers to match England's yeoman class. But the earl had no clear sense of the task before him and no vision of the society that was his to construct. If anything, he took a modern landlord's view of his role, behaving in a manner that worked to the good of the proprietorship narrowly conceived, as was soon made evident in his land-tenuring policies.

It was not entirely the shortsightedness of the earl that allowed a small plantocracy to rise to unchallenged heights in Barbados. In the short period between the settling of the island by the Powell Company and the time he secured his patent, plantation owners were already growing accustomed to managing their households without external interference. The earl would have met with determined opposition had he seriously tried to reduce their authority. Charles I further weakened Carlisle's position by carelessly granting the island to a second patentee.

Seven months after Carlisle obtained his patent (February 1628), perhaps out of ignorance of the area involved, the king conceded to the earl of Pembroke and Montgomery several islands in the West Indies to be called "Provinica Montogomeria."[35] This patent granted to Pembroke proprietorship of Trinidad, Tobago, Barbados, and the nonexistent island of St. Bernado. Realizing that Pembroke's grant raised challenges to his ownership of Barbados, Carlisle informed the king of his error, and thus he was able to get a second patent (April 1, 1628) that made it explicitly clear that Barbados was to be included in his grant.[36] But the damage had already been done. The two earls were soon vying for the favors of planters on the island, which only strengthened the hand of the burgeoning plantocracy.

Exercising his legal rights, in the spring of 1628 Carlisle gave several merchants in England permission to settle a plantation on Barbados.[37] To oversee their estate the merchants chose a former Bermudian colonist, Charles Wolverston, to be their chief agent. Wolverston's commission, which was restricted to the boundaries of his employers' land, and signed by Carlisle, is illustrative of the extensive powers that Barbadian plantation owners were assuming over their bondsmen. Wolverston was empowered to keep His Majesty's peace, resolve disputes, and punish all lawbreakers on the merchants' estate. For those who went to work under the agent, he was to be their employer, governor, sheriff, judge, jury, and if need be executioner.[38] The squatters may not have had Wolverston's legal right to such broad authority, but on the Caribbean frontier they all claimed similar prerogatives in their households.

While Wolverston went to Barbados ostensibly as an agent of the merchants, Carlisle had to be concerned about bringing the independent plantations under his authority, and there is some possibility that he might have offered Wolverston the governorship of the whole colony if he could get the squatters to recognize him as the lord proprietor.

In April 1628, Wolverston sailed to Barbados accompanied by a party of 80 servants. His landing in June was apparently attended with little difficulty, for a short time after reaching the island he set his people to work on the merchants' plantation. Later that same year the merchants sent an additional 42 men, putting Wolverston in charge of 112 servants, by far the largest work force on the island.[39]

Wolverston was not a man to miss an opportunity for advancement, and he quickly went about the business of building himself a faction. In less than three months he convinced many of the plantation owners that they should renounce their loyalty to Powell and elect him as their governor. While the details of Wolverston's political machinations are lacking, it is not hard to construct a broad scenario. By using the "color" of the earl's letter, which declared Carlisle's possession of the colony, Wolverston probably insinuated the idea of a friendship with the nobleman, who had become the foundation of all privileges and riches on the island.[40] As the governor of the merchants' plantation, he could speak at great length about the broad powers that he enjoyed, suggesting that the other estate owners could expect little interference from Carlisle in the management of their plantations. Powell, on the other hand, had nothing with which to compete against the pretensions of his rival. He had no title to the land, no offices to grant, nothing by which to entertain a faction. The proposition before the squatters was one easily decided: remain loyal to one who could not provide or swear allegiance to a bright and rising star on the horizon.

By September 4, 1628, Wolverston had been elected governor by a majority of the squatters.[41] On that same day it was thought fitting by all those assembled that Capt. William Deane, Richard Leonard, Henry Winthrop, and nine others should be appointed assistants to Wolverston. They were to "execute all such office and offices as may appertain to a justice of the peace."[42] Since Carlisle granted only one plantation on the island, in all likelihood the twelve justices of the peace were independent plantation owners who had forsaken Powell for Wolverston. At least it is plain that Deane, Leonard, and Winthrop were squatters.[43] Wolverston had no authority from Carlisle to appoint justices of the peace, and one must assume that they were empty titles, but the squatters took them seriously. In the fall of 1628 there were not more than twenty plantations on the island—twenty plantations, twelve justices of the peace, and a governor. The jurisdiction of the justices probably did not extend much beyond the limits of their own estates, with the Powellites being the only plantation owners who were not justices.

Unaware of how events were turning in Barbados, in August 1628 Carlisle sent out two commissioners, George Mole and Godfrey Havercampe, to establish a proprietary government in the colony.[44] By the time they arrived, Wolverston had most householders well prepared to receive Carlisle as the lord proprietor of the island.

Shortly after landing in Barbados, the commissioners called all the planters together and appointed Wolverston governor of the colony. Of more lasting importance for Barbadian history, they had the colonists acknowledge Carlisle to be the "lord and owner of Barbados," and to agree to pay the earl as a rent "the twentieth part of all profits arising and accruing in the island," and "all other duties, customs, and payments which did or would grow due" on the island.[45]

Planters were to hold their lands as long as they paid their rents and all the other duties that might be demanded by the proprietor. In other words, their rents were not "certain." Carlisle could increase the exactions at his pleasure, and the tenants had no legal recourse. Both large and small planters under the proprietorship were little more than tenants at will. Naturally, the colonists were unhappy with this arrangement, and some generous act of good lordship on the part of Carlisle could have rectified the situation, but it is a testimony to the insensitivity and narrow-mindedness of the proprietors that they were never able to reach an accommodation with the planters. What emerged between landlord and tenants was an adversarial relationship that made it impossible for the proprietors to provide respected leadership in the colony. Understanding that they had completed their task, Mole and Havercampe sailed away, leaving the colony on the verge of a serious conflict.

Although a majority of the squatters acknowledged Carlisle's proprietorship, not all were willing to submit to the rule of the earl and his governor. A few, led by the original members of the Powell Company, formed an opposition faction. They wooed other planters to their side by persuading them that Pembroke was the rightful proprietary of the colony and, no doubt, by offering more secure rents than Carlisle had given. They soon overthrew Hay's government and deported Wolverston. It was not until after the intervention of the king that Carlisle was able to reestablish control over the island. He immediately replaced Wolverston as the governor of the colony with Sir William Tufton.[46]

Tufton's governorship over Barbados was a brief but nonetheless important period in the island's history. He tried to bring strong central government to the island and, in the process, reduce some of the authority that householders were exercising over servants. If he had gotten support from the earl, he might have succeeded. Plantation owners, for their part, had grown accustomed to dominating the island's government, and they viewed their households as their own private domain where they alone made and enforced rules. They mounted a determined effort to resist Tufton's campaign to bring strong central authority to Barbados.

Not long after his arrival in Barbados in September 1629, Tufton ordered a survey of all lands that had been granted previously and confirmed them under his hand. He also divided the island into six parishes. Sir William sparked controversy by refusing to appoint a council from among the principal plantation owners, as Wolverston had done. He also undertook a bold step to halt cruelties against servants by removing maltreated servants from their masters and placing them with other plantation owners without compensation being made to their former owners. Since the beginning of the settlement, planters ruled their servants with little external control, and there is no question that under these circumstances some masters were guilty of ruthlessly oppressing their bondsmen. By intervening on behalf of the servants, Tufton was in effect reducing the authority of planters on their land, and he also raised suspicions that he was trying to mold these abused servants into a force that he might use to diminish the influence of plantation owners.[47]

Tufton's actions produced a storm of protest. Carlisle received complaints that Sir William treasonably asserted his right to be governor for four years "in spite of king, council, or Lord of Carlisle," and that he autocratically opposed all the rules and laws of government that developed in the colony. If Carlisle had supported Tufton at this critical juncture, the plantocracy might have been cut off from power, but Carlisle sided with the planters. On March 15, 1630, he commissioned Henry Hawley to be the new governor of Barbados. Privately, Carlisle instructed Hawley to depose Tufton by force if the situation warranted it.[48]

The new governor arrived in the colony on June 30. He presented his commission to Tufton, who meekly surrendered the government. Sir William, however, must have been beside himself with anger to see that planters had succeeded in removing him from power. Thus, rather than leave the colony, he patiently waited for an opportunity to revenge himself. An occasion soon presented itself. The year 1630 was a drought year in Barbados, and the islanders were in dire need of victuals. When a ship arrived from England with provisions for sale, Governor Hawley forbade anyone to come to the ship without his license. Spreading fears that Hawley would not deal evenhandedly with the planters, Tufton mustered thirty armed men to descend upon the governor's plantation. This assault was more than a food riot, as was revealed in Sir William's proclamation of freedom to all indentured servants that joined him in the revolt. It seems that Tufton was still calculating that the support of maltreated bondsmen would allow him to dominate the plantation owners. His plan miscarried, however. Hawley managed to escape the attack upon his estate and gathered sufficient backing from nearby planters to subdue Tufton. In appealing to servants, Sir William set a dangerous precedent, and it necessitated a sharp response. Hawley had Tufton and the other rebels brought to trial and executed.[49]

Sir William's misadventures produced no change in the living conditions of Barbadian bondsmen. They continued to be used, as John Fincham—a close associate of Tufton who did not join the revolt—observed, "more like slaves than servants," and future governors were warned by Tufton's unhappy experiences what could happen if they challenged the status quo.[50]

After the execution of Tufton, Hawley was able to bring some semblance of political stability to Barbados. Learning from the calamity that befell his predecessor, and obeying orders from Carlisle, Hawley immediately appointed an advisory council consisting of twelve influential planters. Three of those whom he appointed (Edmund Read, Thomas Ellis, and Reynold Allyne) were to dominate the island's political history for the next two decades. While the Council was subordinate to Hawley, the fact that its members were all big planters meant that the general interest of large plantation owners was the controlling influence in the colony's government. The unrepresented masses—tenants, small planters, servants, and slaves—had no say in the functioning of the government.

By 1630, three years after the first settlers arrived, the heads of the largest households on the island had formed themselves into a small plantocracy. Collectively, they owned most of the island's arable land and held on their plantations nearly half of its population as servants and slaves. For all intents and purposes, no force (church, state, or community) regulated how they treated the bondsmen in their households.

CHAPTER 3

The Tobacco Era

Looking back on the Barbadian settlement in the 1630s, only the most wildly optimistic settler would have predicted prosperity for the island in the next decade. Unlike the rather stately houses, encircled by acres of well-cultivated lands that the sugar plantocracy enjoyed, Henry Colt described early tobacco plantations as resembling "the ruins of a village lately burned—here a great timber tree half burned, in other places a rafter all black." No plantation, he complained, was "weeded for beauty; all are bushes, and long grass, all things carrying the face of a desolate and disorderly show to the beholder."[1]

In the English Americas, before the sugar revolution in Barbados, those who sought to sell agricultural products to a European market tended to concentrate on tobacco cultivation. The success that Virginia had with the weed in the 1620s spurred on efforts to organize tobacco plantations in the Caribbean.[2] Despite the high expectations in Barbados, planters there soon learned that the island's soil produced a tobacco that was barely worth exporting. Henry Winthrop's aspirations for making himself a fortune on the island were greatly chilled when he was abruptly informed by his father that the tobacco he sent home was "very ill-conditioned, foul, full of stalks and evil coloured."[3] The English market believed that Barbadian tobacco was "foul and had an earthy taste"; consequently, it did not bring the best prices in the mother country.[4]

Besides economic dislocation, the tobacco era in Barbados was also known as a period of widespread political unrest. The commissions issued by Carlisle to his governors were devoid of institutional definition. Carlisle's first governor, Charles Wolverston, selected a Council of twenty assistants to help him rule. It would seem that Wolverston was not instructed by Carlisle to appoint a council. He probably thought that it would be politically expedient to have a council to advise him. His successor, Sir William Tufton, tried to govern without a council to his undoing. Except for Tufton's administration, until 1638, the island's government consisted of a governor, his personally appointed Council, a receiver of the rents, a secretary of state (who filed deeds), and locally elected vestries that were also subordinate to the governor.

During this period, the governors and councils held supreme command. Councillors remained subservient to the governor who was appointed, but their participation in his acts lent a less arbitrary guise to the administration. In 1638, rebellious planters organized a popularly elected Assembly which the proprietors were ultimately forced to recognize as a part of the island's government.

The relationship between the governor, the Council, and the Assembly (together they formed the island's General Assembly) is difficult to recapture, as the evidence is meagre. In the minds of most Barbadian planters, the Assembly functioned as a miniature House of Commons. The Assembly represented the popular element in the government. It refused to sit with the Council, seeing the latter body as the island's equivalent of the House of Lords. As the colonists insisted on many occasions, no taxes could be imposed upon them without the consent of the Assembly. In the General Assembly, the governor, the Council, or the Assembly could veto any bill. The governor, with the advice of the Council, acted as the executive branch of the colony's government.

The political community was not very large. In the main, it consisted of adult white male heads of households; the biggest estate owners dominated local affairs. Sovereignty in the colony did not reside in the local political community but in a distant metropolis in the person of the English king. Whatever disadvantages this distance between ruler and ruled might have had for political order on the island, the crown enhanced the prospects for colonial unrest by delegating its authority in the form of a proprietary grant to Carlisle. This transfer of power further attentuated the bond between ruler and subject, with the result that whenever questions occurred about legal rights to the proprietorship (to cite an example that had repeated relevance), they fostered factions in the colony as householders sided with that landlord who best protected their interest. The royal grant also conferred upon the earl the right to appoint all officeholders in the society. Barbadian magistrates, consequently, were not directly responsible to the island's inhabitants; rather, they served at the pleasure of an absentee landlord who was blind to their conduct in the colony by distance and time.

These structural weaknesses in the island's political system were exaggerated by the problem surrounding communication and implementation of instructions from the mother country. Communication between Barbados and England was slow and irregular; hence, directives from England were generally out of touch with the rapidly changing political reality of the island. Complicating matters, Barbadian factions sent out one-sided information that created uncertainty, making it difficult to determine the "legitimate" state of political affairs in the colony. If the interest of English local communities could sometimes deviate from that of the central authority, it is not surprising that on the West Indian frontier there was an even stronger proclivity for the material aspiration of Barbadian householders to diverge from the mercantilistic policies of the metropolis, and in particular from the interest of the absentee landlord. Keenly sensitive regarding their profit, the planters consciously evaded or willfully misinterpreted the spirit as well as the letter of commands from the metropolis when it was to their advantage. In Barbados there was not that singularity of purpose between ruler and subject or willing responsiveness to governmental instructions upon which political tranquility is predicated.

The earl of Carlisle, near death and desiring to repay some of the outstanding debts incurred in the settling of Barbados, in the spring of 1636 put his West Indian possession in trust under the care of Sir Archibald and Sir James Hay, his cousins, and Richard Hurst (Hurst died shortly after his appointment). They were to oversee the island and use the profits received thereby to reimburse the earl's creditors.[5]

The trustees took a genuine interest in Barbados and attempted to place the colony on more secure footing. When they took over the proprietorship of Barbados, there were just over 2,000 people on the island, but in 1638 the figure had grown to nearly 9,000.[6] While they were more attentive about encouraging the development of the colony, their interest sprang from a desire to expand the size of the profits that could be claimed from the colony. They proved to be landlords in the modern as opposed to the feudal sense of the word.

In keeping with their chief concern, the trustees sent their cousin, Peter Hay, to receive all the rents, profits, customs, and duties owing to the proprietary estate in Barbados. Hay was also instructed to seek out all arrears and to give a continuous report on all matters affecting the profit and rights of the proprietorship. Of great importance, Hay was not to be a subordinate of the governor. Rather, he was answerable only to the trustees.[7] While Receiver James Holdip, Peter's predecessor, also enjoyed such autonomy, he seems to have allied himself with Hawley, and he looked away as the governor assumed possession of goods and lands that belonged to the proprietorship. With ambitions of rising above the governor, the new receiver made a more determined effort to collect all that was due the proprietor, including that which was previously misappropriated by the governor. His inquiries into Hawley's past misdeeds sparked a political crisis in Barbados.

Hay reached the colony in September 1636 and was heartily welcomed by the governor. He wrote to his employers that Hawley promised him "all the assistance as lay in his power." Peter characterized the governor as "very diligent in doing all things for the good and profit of the island," but he added in a distrustful tone that "as for any concealed things I cannot find none, not yet."[8] During this get-acquainted period, the governor and the receiver remained on friendly terms. In fact, Hawley appointed Hay a member of his Council.

As time went on, however, relations between the two became strained as the receiver began insisting that certain revenues that the governor formerly appropriated to himself were rightfully owed to the proprietor. Taking advantage of Hawley's departure to England in 1638, Hay presented eleven demands to Deputy Governor William Hawley, the governor's brother. Nine of the articles he presented concerned money and goods the governor had received that Hay claimed properly belonged to the estate of Carlisle. They included customs, revenue from the granting of land, the session house that the first earl built for the island's use but was now being employed for the private benefit of the governor, entry fees collected from newly settled householders at the rate of 100 pounds of cotton or tobacco for every member of their household, and the forfeitures from the estates of felons. The Council responded to his demands by telling Hay that he should not meddle with anything that was due "unto my Lord, but only the twenty pounds per head." Moreover, they added menacingly, "the governor had a great deal of patience in suffering with him for so long." Angry with Hay, William Hawley removed him from the Council.[9]

Complicating the political situation in Barbados was the maneuvering of James Hay, second earl of Carlisle, who was trying to wrest control of the proprietary patent away from the trustees. Unhappy with his exclusion from his father's West Indian holdings, Carlisle proposed to the fiduciaries that he take over the trust, but they refused his offer. Their rejection apparently did not deter the earl. In October 1637 the trustees informed Peter Hay that "there hath been some differences between my Lord Carlisle and us, but we have submitted [it] to friends, and are like to agree." These negotiations did not conclude to the satisfaction of Carlisle. In 1638 he presented his case to the king, but once again his plans to assume control of the colonies failed.[10] Carlisle was never able to break his father's trust, though his continued interference in Barbadian affairs did contribute to the destruction of proprietary government in the colony.

The situation was further confused by the proposal of Robert Rich, second earl of Warwick, to purchase Barbados. In late 1638 or early 1639, Warwick bought the Pembroke patent with its dubious claim to Barbados. Unfortunately, there is no way of telling if he acquired the patent before offering to buy Barbados from Carlisle and the trustees. Apparently, Warwick tried to persuade the proprietors that by selling the island to him they could settle the deceased earl's debts, and the younger Carlisle could take possession of the other islands granted to his father. Carlisle seems to have been willing to accept Warwick's proposal, but the trustees were opposed to it.[11]

The first word of Warwick's offer reached Peter Hay in October 1638 when he was notified by Carlisle that he was to be accountable to Warwick for the proprietary rents. Worried about his future, Hay wrote the trustees that he would not recognize Warwick's ownership of the island without written instructions from them.[12] Shortly after Hay sent his letter to his cousins, Warwick sent a message to the colonists that he expected soon to conclude his purchase of Barbados but that the transactions had been delayed "through the interposing of some uncertainties" by the trustees. Once these "inconsiderable differences" were settled, he promised to confirm the existing government to the extent that it accorded "with order, civility, and justice." As for Governor Hawley, who was still in England, Warwick assured them that "he will be accommodated to his contentment."[13]

The new indeterminacy about who owned Barbados created divisions in the colony as householders rallied behind the potential proprietor that best advanced their individual interest. In a 1639 letter to the trustees Peter Hay left some indication of the nature and extent of the factionalism on the island. "The governor and the most part of the council," he surmised, were for Carlisle. They feared that a change of ownership, Warwick's ambiguous overtures notwithstanding, would mean their dismissal from office; and, given the demands of Peter Hay for an account of past revenues, they had little confidence in the trustees. A second faction grew up around Receiver Hay. Sharing similar apprehensions with the governor and the Council about what a Warwick proprietorship might mean for their tenure in office, Hay remained loyal to the trustees.[14]

"The country," Hay believed, was "desirous to have Warwick to be their lord in this place."[15] Aware of the general unrest on the island provoked by disputes with the proprietors over rents and land tenure, and sensible that many householders, particularly small planters, were disturbed by the rampant corruption of Hawley's government, Warwick began a vigorous campaign to create a sizable following in Barbados through whom he hoped to encourage the sale of the colony. He let it be understood that he would reform the government and establish a more equitable system of land tenure. Weary with the abuses and uncertainties of the Carlislian proprietorship, the majority of the colonists flocked to Warwick's standard. The islanders expected the earl to improve the system of justice "that doth make laws one court and break them the next."[16]

The leading supporter of Warwick in the colony was James Futter. Futter was one of the original squatters. He refused to consent to the agreement that Mole and Havercampe obtained from the squatters. Throughout the 1630s, he remained a thorn in the sides of the proprietors, challenging their rule as an illegal imposition. Futter argued that as the squatters settled the island before the proprietary patent was granted to Carlisle, they should enjoy their property without acknowledging him as their landlord or paying any rents to the earl. Sometime around 1638 Futter became Warwick's agent in the colony. Besides continuing his denunciation of the Carlislian proprietorship, he attacked the character of the earl and the Council. He charged that Edmund Read, one of Hawley's councilors, had fathered a bastard. When dragged before the Council to answer for his remarks, Futter boldly asked, "if all whoremasters were taken off the bench, what would the governor do for a council?"[17] He further stated in open court that "my Lord Carlisle himself was somewhat too much given to drink." For his impudence Futter was ordered to stand in the pillory for "an hour in [a] violent hot day, without his hat, it being then so parching hot that the sun pierced his skull."[18]

Barbadians were so impressed with Warwick's "fair promises" and with the intrepid actions of Futter, Peter Hay told the trustees, that the "country is desirous to have Warwick to be their Lord in this place." Indeed, most heads of households were so confident that Warwick would buy the island, the receiver wrote, "that they are very backward in paying of their rents that I have not received above ten thousand pounds of goods all this year." "They have come to such a pass," he reported, "that they will not obey the governor's warrant or attachments."[19] While there was considerable enthusiasm on the island for Warwick to become proprietor, in different times the earl would not command such a following.

Matters reached a head in the spring of 1639. The trustees told Peter Hay and the deputy governor that the bargaining with Warwick had been terminated. Although there were a few planters, James Futter in particular, who continued as loyal supporters of Warwick, the news that the earl would not be purchasing the island destroyed the Warwick faction. The trustees also announced that a compromise had been reached with Carlisle to make Sgt. Maj. Henry Huncks governor of the island. They prophetically warned the colonists that Hawley might return to the colony before the new governor arrived and attempt to make a "party against Huncks." In such an eventuality, they advised the colonists that neither Huncks nor Hawley should be recognized as governor but that the deputy governor should continue in office till further directions were received from the metropolis.[20]

Carlisle also anticipated that Hawley would not accept his dismissal peaceably. He apparently suspected that Hawley would collaborate with a rejected Warwick to entice Barbadian householders (who had already expressed affection for Warwick) to migrate to one of the islands that the earl purchased from Pembroke. Attempting to prevent a disastrous exodus from his colony, Carlisle persuaded the king to write a letter to Barbados on March 16, 1639. In his letter, Charles ordered that any person who sought to seduce anyone to leave the island should be resisted by all lawful means and that no person should be permitted to depart the colony without permission. Furthermore, Charles demanded that all inhabitants submit to Huncks.[21]

While Carlisle was wrong in thinking that Warwick and Hawley might be scheming to depopulate his island, he was right in suspecting that Hawley would not step down from power without a contest. The deposed governor was abetted in his efforts by a serious blunder on the part of the crown. On March 27, 1639, Hawley received a commission from the king to contact the several colonial governors about the cultivation and marketing of tobacco in America, in hopes of ending overproduction and raising the price of the staple. The commission mistakenly described Hawley as "lieutenant general and governor of the island of Barbados."[22] In Barbados, he used this error as a smokescreen to cover his plans to retain his office.

On the day that Charles marched northward against his rebellious Scottish subjects, Hawley left England for Barbados, arriving in the colony in late May. In a matter of weeks, he was able to foment a full-scale revolt against the proprietorship. Under the cover of the king's letter, he replaced all proprietary officeholders with his own people. He also ordered the islanders to stop paying their rents and freely handed out lands belonging to the Carlisle estate. Of more far-reaching importance, Hawley illegally called for the electing of burgesses for a general assembly. Two representatives from each of the parishes were to be selected; every planter who held ten acres of freehold land was eligible to vote.[23] (Technically, no settler had freehold land in Barbados, but apparently anyone who held their lands directly through the earl was considered by the islanders to be a freeholder.) In this bold stroke, Hawley enfranchised about 1,500 Barbadian planters, half of whom were large plantation owners. The sizable tenant population and small planters with less than the requisite ten acres were not given the vote. From the beginning, the Assembly was a creature of the plantocracy. The members of Hawley's lower house were some of the biggest landholders on the island. Before the proprietors could put down this revolt, they were forced to recognize the Assembly as a legitimate part of the colony's government. In time, this body would lead a more successful rebellion against the Carlislian proprietorship.

With Hawley liberally giving out offices and land, and enfranchising the plantocracy, even those who earlier supported Warwick in opposition to Hawley readily joined the displaced governor's camp. When Huncks reached Barbados, he found the island's political community unwilling to see him sit as the colony's chief magistrate. Shortly after his arrival, he went before the General Assembly to present his commission and the king's letter (that of March 1639)—both of which Hawley and his followers "did all extremely slight."[24] In words indicative of the mood that planters were in, one member of the lower house proclaimed that the "island did belong to the king of Spain, and not to the king of England." He went on to threaten that "if the Earl of Carlisle or any other lord of England should come to take any of his goods from him (the question being the usual duties to the earl), or should stop him on the island he would cut his throat."[25] A member of the Council echoed the burgess' inflammatory language. He asserted that "he had more authority on the island with his staff or rod, which he bore in his hand, than the King of England."[26] Hawley, according to Huncks, held both of these men in a "special trust." Realizing that there was little good to be done with men who had gone so far as to doubt the authority of the English king, Huncks left the island and wrote to Carlisle of his experience.

In the fall of 1639, Carlisle and the trustees joined together in a petition to the crown for assistance in deposing Hawley's illegal government. On December 22 the king appointed Peter Hay (who fled the colony shortly after his foe, Hawley, returned) and three others to displace Hawley and to turn the government over to Huncks.[27]

The commissioners landed in Barbados in late March or early April 1640. They took Hawley into custody and forced him to sign a declaration formally resigning the governorship.[28] Unable to further resist commands from the mother country, the plantocracy quietly permitted Huncks to be installed as governor.

With large planters already hostile toward the proprietorship, Henry Huncks was a particularly unfortunate choice to replace Hawley. He was overly quarrelsome, given to strong drink, and too inept to handle the subtle business of colonial politics. Further handicapping the new governor was the fact that he operated without the genuine trust of Archibald and James Hay. He was essentially Carlisle's governor.

One year after assuming power in Barbados the island's plantocracy was on the verge of open rebellion against Huncks's rule. Huncks's problems began with his decision not to bring large planters into his government. Instead, he appointed "strangers who he brought with him," and a number of planters of dubious reputation on the island (e.g., the profligate son of Sir Edmund Verney, Thomas). A cry went up in the colony that these new public officials would not be concerned with the welfare of the settlement but would be out for "their own private gain."[29]

Unlike Tufton, it does not appear that Huncks made any efforts to acquire the loyalty of servants or small planters as a counterforce against the influence of large householders. He excluded the latter from power because they had given in to Hawley's shallow deception, and he wanted revenge.[30]

The new governor was not entirely at fault for the political crisis that engulfed Barbados during his administration. The conflict between Carlisle and the trustees greatly contributed to the island's instability. In spite of his promises to the trustees, Carlisle quietly instructed Huncks to change the method of collecting the rents by establishing a rate of three pounds of cotton or tobacco per acre instead of the forty pounds per poll. The earl also empowered the governor to audit the accounts of Receiver Hay. His purpose was clearly to take over the collections of the rents from the trustees.

Following Carlisle's instructions, one month after his installation in office Huncks told Peter Hay that he had authority to call him to account. The receiver immediately wrote the trustees of the danger of subordinating him to Huncks. If he gets the estate into his hands, Hay warned, "you shall not blame me for it." He was confident that if Huncks were able to get the collection of the rents under his supervision the trustees would have a small account of what was due them.[31]

With the rents a possible bone of contention between the fiduciaries and Carlisle, Hay saw an opportunity for himself. He proposed some alterations in the way the proprietary rents were distributed that would have left Huncks without control of the public budget. The previous governor, Hawley, was allowed one half of the forty pounds of tobacco per poll from which he was to purchase ammunition, fortify the island, and pay for whatever else was necessary for the functioning of the government. After these expenditures were made, the governor was permitted to keep the remainder as his salary. Hay recommended that Huncks be given a personal allotment that would not be burdened with public expenses.[32] The implication was that the receiver would become the public paymaster on the island, thereby increasing his authority over the governor.

While Hay advanced his proposals to the trustees, Huncks, on the behalf of Carlisle, reached an agreement with the Council and the Assembly for the resettling of the proprietary rents. Under this new system tenants were to pay three pounds of tobacco or cotton per acre of land. In addition, Carlisle was to have two and one half pounds from the buyer and an equal amount from the seller of land in Barbados. The governor was to have one half of the rents collected, promising to meet all public charges from his share of the revenue. Declaring that Carlisle would not ask for a larger rent, Huncks granted tenants "confirmation of their lands to them and their heirs forever to be held by knight service."[33]

There was tremendous confusion among householders over which rental system would work to their best advantage. Hay spread rumors that the governor tricked the General Assembly into an agreement that would increase the amount of money that the islanders paid the proprietor. Hay also told planters that by the new system they were "tenants-of-will, or for seven years at most."[34] Since the type of knight service was not specified in this accord, planters did not gain any greater security from a capricious act of the proprietor, and Hay's words had an unsettling ring of truth about them. There were also complaints about the new charges placed upon the sale of land.

Huncks heightened suspicions that some scheme was afoot to defraud the colonists by stripping William Powry, a member of Hay's trustee faction, of the profits and most of the responsibility of the secretary of state. Powry was no longer to keep the records of land grants and sales. Huncks vested these powers in an alienation office that was operated by persons appointed by him. As Powry was appointed by Archibald Hay, and the trustees viewed Huncks as Carlisle's agent, they interpreted the governor's actions as an attempt to diminish the importance of their commission. In the colony, Hay provocatively informed the colonists that the changes were made to conceal the profits that were being reaped by the new rental system.[35]

Given the uncertainty about what effect the alterations in the rents might have, coupled by the fact that most influential planters were opposed to Huncks, it is not surprising that a public outburst against the governor occurred. In February 1641, the vestry of Christ Church drew up a petition to present to Huncks. The gist of the vestry's complaints is probably in their supplication to the assemblymen from that parish, which was apparently written at the same time as the petition. Among other things, the burgesses were asked to take care to remedy all grievances of any man who suffered either in his person or in his estate by the hands of any magistrate "who had not duly administered justice unto him," and the offending justice be removed from office. It was also requested that in the business of the rents the assemblymen be circumspect and not alter the system "without it be for the country's advantage and reputation." Furthermore, it was entreated that "places of command, military, and civil, may be conferred on men of both estate and judgement," and that special care should be taken that "none be continued in office which are not furnished with neither of those benefits of virtue and fortune."[36] In the latter clause, the Christ Church vestry was demanding that only the island's plantocracy be in a position to govern.

The petitioners marched from their parish church to within one-half mile of the governor's house. From there they dispatched a messenger to inquire whether Hunсks would receive them. The governor arrested their messenger. When they failed to get a reply, the crowd moved closer to the governor's plantation and sent two more of their company with the petition. Peter Hay met these delegates on the road and joined their party. They found Huncks armed and threatening that "if the gentlemen had come up to his house altogether, he would pistol the first of them and hang the rest." Hay recommended that Huncks receive their petition, but the governor refused. Instead, he ordered his provost marshal to take into custody all those that were in a "mutinous manner drawn together"; those arrested were detained in prison for nine days.[37]

Expecting the imprisonments to "raise a great deal of trouble in the country," Receiver Hay stoked the fires of disorder by issuing a warrant for the collection of the rents.[38] He demanded that it be done according to the old way, forty pounds of tobacco per poll. Huncks reacted to this provocation by ordering Hay arrested, but the receiver was released after a short stay in jail. Recognizing now that the planters had concluded that the new rental system would increase what they paid for their lands, the governor tried to placate them by agreeing to return to the forty pounds per poll.[39] Before his initiation could take effect, the proprietors recalled the governor.

The proprietors attempted to restore peace to the troubled island by removing both Huncks and Hay from office in June 1641. The trustees explained to Hay that Carlisle would never have consented to recall Huncks if they had not agreed to appoint a new receiver general. The trustees might have done more to protect their servant had they not fallen into a bitter dispute with Hay over his financial accounts. They were also disturbed by reports that Hay raised a commotion in the colony by suggesting that he was the legal heir of Carlisle (unless the earl had children), and hence would someday inherit the island.[40] The landlords named James Browne, a kinsman, receiver of the rents. Unable to agree upon a permanent replacement for Huncks, Carlisle and the trustees compromised by selecting Philip Bell to be deputy governor of the colony.[41]

Many contemporary observers of Barbadian society were of the opinion that the constant political turmoil that beset the island in the tobacco era was in great measure responsible for the colony's slow economic development.[42] Certainly the feuding between Powell and Wolverston, Hawley and Huncks, and proprietors and tenants had their ill effects; but more closely related to the problems of economic growth was the fact that Barbados could not produce more than an inferior grade of tobacco.

The situation became more critical in 1631, when the English Privy Council set a quota on the output of tobacco to be allowed from Barbados, ostensibly to encourage the colonists to grow other crops and ease a developing glut in the home market. No tobacco other than that which was "sweet, wholesome, and well packed" was to be exported to England. Thereafter, the Privy Council passed a number of measures regulating the tobacco trade. Barbadian tobacco paid one shilling duty per pound; Virginia had a levy of only nine pence against it; and Spanish leaf, as it originated outside the mercantile empire of England, had to face dues of two shillings a pound.[43]

The duties were soon reduced, but they, with the restrictions on quality, were enough to encourage Barbadian planters to experiment with other crops. In the early 1630s a number of householders cultivated grain on their lands, the principal one being Indian corn. Father Andrew White, visiting Barbados in 1634, described the island as "the granary of all of the Caribbee Islands."[44] Despite White's high praises, the physical environment, as with tobacco, inhibited the production of grain. Passing a Barbadian cornfield, the perceptive Henry Colt noticed, "it was most part of it blasted." Colt believed that the crop was in this state "because their plantations are so near the seashores and by that means made subject to the north and east winds which continually blow."[45] In addition to environmental problems, there was no ready market in the Caribbean for the island's corn.

In 1631 Henry Colt wrote in his diary that "now the trade of cotton fills them with hopes."[46] By 1634 Father White could list cotton as another of the staples of the colony. The farming of cotton, however, did not progress very far. In October 1637 the proprietors informed tenants who persisted in tobacco cultivation that "Barbados tobacco cannot expect to come to a good market anywhere, it hath the reputation to be so bad." They therefore recommended to householders that they should plant cotton, "which is a staple commodity, and had the reputation to be better in Barbados than any place of the world."[47] In spite of encouragement from the landlords, cotton did not become the dominant crop in Barbados. Peter Hay, who rarely had a good word to say about tenants, told the proprietors that the heads of households in Barbados would not cultivate cotton, "because they are so indebted that if they leave the planting of tobacco they shall never be able to pay their debts."[48] Daniel Fletcher, one of the planters commissioned to restore peace to the island after Hawley's rebellion, gave the trustees a more plausible explanation for the general refusal to plant cotton. "The most part of the country," he informed them, "will bear no cotton."[49] It was established that only "one acre in ten" was suitable for cotton cultivation.[50]

The economic condition worsened in Barbados after the fall of tobacco prices in 1637. With settlers in Virginia, St. Kitts, Providence Island, Bermuda, and Barbados shipping over 1 million pounds of tobacco to England annually, the English market became severely glutted. The oversupply caused Barbados's earthy weed to plummet in value. On discovering that Barbadian "tobacco would not yield any money in England," Peter Hay decided to ship it to Holland.[51] Avoiding the saturated market of the mother country, however, did not improve the price merchants were willing to pay for it.

As the colony's major export lost its worth, English merchants began to frequent the island less and less, leaving the colonists with a dwindling supply of goods and credit from the mother country. Indeed, had it not been for the willingness of Dutch traders to stop at the colony on their way to their own possessions in the Caribbean, the Barbadian settlement might have miscarried.[52]

Desperately struggling to increase the returns from tobacco, in 1638 the islanders joined with planters in several of the other English colonies in declaring a two-year moratorium on the planting of the staple in hopes of ending the surplus in the metropolis.[53] As a consequence of this suspension, Barbadian tobacco imports in the port of London dropped from 204,956 pounds in 1638 to 66,895 pounds in 1640.[54] It was during this hiatus that plantation owners began to turn their attention to sugar cultivation.

The tobacco era in Barbados was a time of widespread political disorder and economic depression. Sugar and slaves solved the economic difficulties, but they also brought a host of new political problems that exacerbated the old feuds that had developed in the 1620s and 1630s.

CHAPTER 4

Sugar and Slaves

At the precise moment that a liberal democratic consensus was struggling to power in England, and during a period of home rule in Barbados, the island was transformed from a servant-tobacco colony to a sugar monoculture whose laboring force consisted mainly of black slaves. According to the computations of Richard Ligon, by the mid-seventeenth century the Barbadian economy could produce in twenty-two months over £6 million in sugar[1]—a volume that attracted trade from as far away as Russia, and provided the New England colonies with the commerce they desperately needed to survive their infancy.[2] In these halcyon days, Barbados was known by Englishmen as the "jewel" of the British Empire.

Barbados was not the first English colony to experiment with sugar. Conscious of the potential riches to be gained from sugar, in the 1620s and 1630s colonists in Bermuda and Providence Island attempted to produce marketable crops without success.[3] In 1638, Archibald Hay told Receiver Hay that the earl of Warwick intended to start a sugar plantation on Barbados.[4] This is the first mention in the sources of an economic interest in sugar, and Warwick might well be responsible for planting the seeds that were to flower as the sugar revolution.

James Holdip and James Drax are usually credited with being the first planters to actually grow sugarcane in the colony. They obtained the plants from Pernambuco, Brazil; to their delight, these plants thrived in Barbados, which thus encouraged them to set up a small sugar works.[5] By 1643, Thomas Robinson, a proprietary agent in St. Kitts, could write home that "Barbados is grown the most flourishing island in all those American parts, and I verily believe in all the world for the producing of sugar."[6] While this glowing report is indicative of the spreading news of the success the island was having, the Barbadian sugar industry was still in its infancy.

It took nearly a decade for planters to learn how to efficiently organize their households for the growing of sugarcanes and the manufacturing of sugar from the cane plants. Initially, the sugar produced in Barbados, Ligon recounted, was "inconsiderable and of little worth."[7] Thomas Peake described it as a "kind of coarse sugar, which we call Barbados-sugar," which will not "keep long, not that the country is unapt for better, but, as 'tis rather supposed, because the planters want either skill or stock to improve things to the best."[8] In later years plantation owners would attribute the quick spoilage of their sugar to the harvesting of the cane before it was fully ripe. Learning from their mistakes, and seeking out the advice of Portuguese planters in Brazil, by the 1650s Barbadians had acquired sufficient skills to manufacture a highly marketable commodity, though in Ligon's view it was still not as good as the sugar obtained in Brazil.[9]

The discovery of two Barbadian customs books for the years 1664–66 inclusive provides an important opportunity to determine the extent to which sugar dominated the island's economy by the mid-seventeenth century.[10] One of these ledgers, "A Coppie Journall Entries Made in the Custom House of Barbados Beginning August the 10th, 1664 and Ending August the 10th, 1665," has been subjected to a computer analysis. The book is 200 folio pages in length and has over 6,000 entries. Before discussing the findings, some general comments on this source would be useful.

In 1663, Barbadians agreed to pay the English crown a 4.5 percent duty on all produce exported from the island.[11] The customs books are a record of the items shipped from the colony that were subject to the duty. They list the date of export, the name of the exporter, the commodity shipped, and its weight (see p. 58). To facilitate the assessing of the duty, after the weight of a particular staple was given,

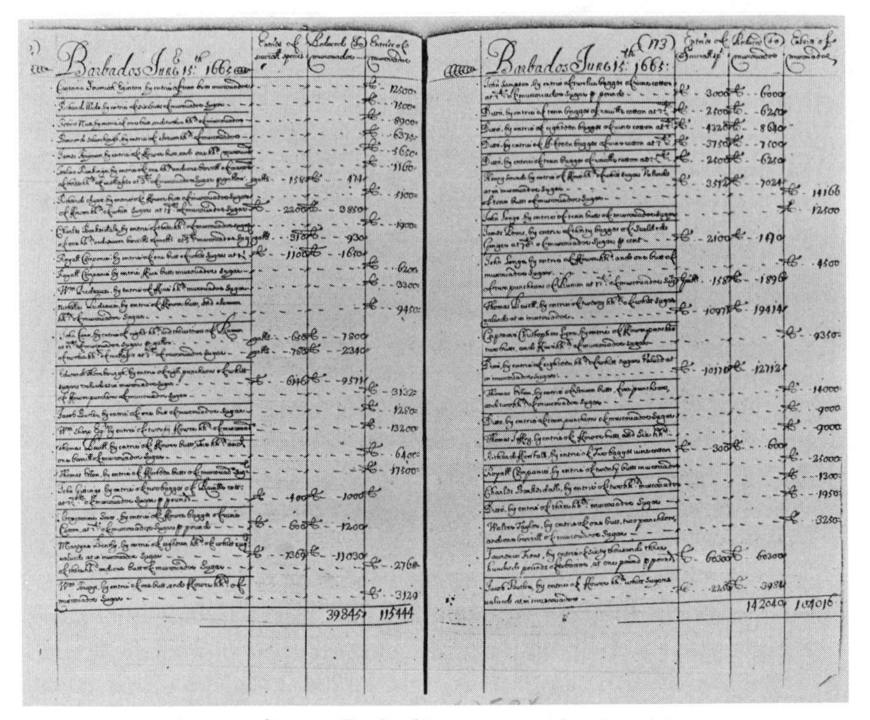

Barbados June 15th 1665

39845 115444

(173)

Barbados June 15th 1665

142040 104016

A page from a Barbadian customs book, 1665.
SOURCE: A Coppie Journall Entries Made in the Custom House of Barbados Beginning August the 10th, 1664 and Ending August the 10th, 1665, Ms. Eng. hist. b. 122 (Bodleian Library).

it was converted to a comparable weight in muscovado sugar. For example, 1 gallon of rum was valued at 12 pounds of muscovado sugar, so that 3 gallons of rum was worth 36 pounds of muscovado sugar. It should be stressed that the assessed value of a commodity, rum at 12 pounds or ginger at .70 pounds of muscovado sugar, were not determined by current market prices but were politically fixed rates that did not vary over the period covered by the two customs records.

It should also be noted that the names appearing in the customs books as the exporters were not necessarily the actual owners of the staples shipped from the colony. Planters did not always accompany their produce to the Bridgetown customshouse, leaving that matter to servants, factors, and agents. In England, it was also common practice for factors to ship goods for their masters, but English customs records noted the name of the agent as well as the rightful owner of the commodity. Unfortunately, in Barbados customs officials did not go to this trouble, so that it is impossible to say whether a person listed as shipping a commodity from the island was sending his own property or someone else's. Consequently, the customs records cannot be used to measure the productivity of individual plantations or the distribution of wealth in Barbados.

It is also impossible to determine how well these records portray the general productivity of the island's economy. As these ledgers were kept for the purposes of collecting a duty, presumably smuggling took place that they did not register. The outbreak of the Second Anglo-Dutch War in the spring of 1665 might have made many potential exporters hesitant about shipping their goods in uncertain waters. In more peaceful times, the volume of the island's exports might have been substantially higher.

For all of its limitations, the computer analysis of the customs book leaves no doubt that sugar was the most important staple in the Barbadian economy. Between August 1664 and August 1665, Barbados exported various commodities valued at over 28 million pounds of muscovado sugar (see Table 4.1). Eighty-seven percent of the items shipped from the colony were a derivative of sugarcanes (rum, molasses, and refined and muscovado sugar). The closest runner-up was cotton, which accounted for only 11 percent of all exports; tobacco was a distant third at .79 percent. Sugar was king in Barbados.

The rise of sugar cultivation brought a frenzied, bonanza climate to the colony. Barbados and sugar became hypnotic words that induced daydreams of great riches and brought shipload after shipload of settlers to the island. They came from England, Ireland, Scotland; a few were from France, the Netherlands, Portugal, and from other parts of the English Americas. As mentioned earlier, Samuel Winthrop thought he would be able to live better in Barbados "than in any other place," though his greatest difficulty would be at his first settling, his "stock being small."[12] Thomas Modyford, who founded a fortune in sugar and became governor of Barbados and then Jamaica, resolved after his initial landing in 1645 in Barbados "not to set his face for England, 'til he made his voyage, and employment there worth him a hundred

TABLE 4.1
Barbadian Exports
August 1664–August 1665

Commodity	*No. entries in journal*	*Value in muscovado sugar*	*Percent of total value exported*
Muscovado sugar	3,034	20,040,524	70
Refined sugar	447	3,399,065	12
Cotton	1,266	3,142,025	11
Rum	551	1,232,922	4
Molasses	300	235,168	.82
Tobacco	37	224,816	.79
Ginger	167	190,881	.67
Indigo	14	23,832	.08
Fustic woods	7	16,850	.06
Miscellaneous	8	7,446	.03
Total	5,831	28,513,529	

Source: A Coppie Journall Entries Made in the Custom House of Barbados Beginning August the 10th, 1664 and Ending August the 10th, 1665, Ms. Eng. hist. b. 122 (Bodleian Library).

thousand pounds sterling."[13] Francis Lord Willoughby, another governor of Barbados (1650–52 and 1663–66), came to the island to "find a way to live." His property had been confiscated "all at one clap" by Parliament for his intriguing with the king. He decided not to "sit down a loser, and be content to see" his wife, children, and "self ruined." Since all was "gone at home," it was "time to provide elsewhere for a being."[14] In the 1640s Barbados was the place to restore lost fortunes, make new ones, and enlarge those already in existence. These were its attractions, but for most the venture turned out to be a fool's errand. Fortunes were indeed to be made in the colony as a result of the sugar revolution, but only a tiny minority were able to accumulate such wealth.

Putting together a sugar plantation was an expensive proposition. Building and equipping a mill, a boiling house, and a curing house did not cost much. The major items of expense were land and labor. As one might suspect, the trend to establish sugar estates created a demand for land that brought an increase in its price. Between 1640 and 1650, the value of ordinary land rose by 200 percent, and the figure was higher for land already planted in sugar.[15] In 1642, William Hilliard's 500-acre plantation could have been purchased for £400; five years later a buyer had to pay £7,000 for only half of it and half of the stock on it.[16] In April 1646, William Powry, a member of the Barbadian Council, reported that this "year a plantation of five hundred acres of land with fifty Negroes and forty men bought and sold for sixteen thousand pounds sterling." He assessed that "there was not an acre of land in the whole island to be purchased for five pounds sterling or the value thereof in commodities."[17]

Richard Ligon estimated that a person coming out to Barbados in order to venture into sugar had to be able to invest £14,000.[18] Unfortunately for those who signed themselves into servitude in exchange for passage to Barbados, they did not appreciate the enormous capital outlay that sugar cultivation demanded. They contracted to labor on the island in the belief that once they served their time land would be granted them, and they would be in a position to begin a sugar plantation—but none of that was to occur. Although land speculators had taken possession of all available land by 1638, in the tobacco era it was possible for a poor man to acquire small tracts of land either by renting or by outright purchase. The sugar revolution so increased the price of land by 1647 that it became far too expensive for those of limited means. Commenting upon this situation, Lord Carlisle wrote that

> divers people have been transported to my island Barbados in America, and have there remained a long time as servants in great labour for the profit of other persons, upon whose account they first were conveighed thither, expecting after their faithful service, according to the convenants agreed upon at their entrance, there to make some advantage to themselves by settling of plantations for their own use, but by reason of the great number of people who repaired thither, and by the blessing of God have multiplied there; the land is now so taken up as there is not any to be had but at great rates too high for the purchase of poor servants.[19]

The sugar revolution also encouraged land consolidation. The two brothers, Stephen and Thomas Noel, appear to have built up a plantation through a process of aggregation. In less than a year, they bought a parcel of ground and house near their land from Charles Ambler; 5 acres from William Snethwick; a "plantation of land" adjoining their own land from William Ayme; 20 acres from Richard Hilliard; 6.5

acres from John Gibson; and 25 acres from John Williams. Since all these purchases were made in the same parish, we may reasonably suppose that the Noels were building up a plantation.[20] William Hilliard gave his half of Kendall's plantation to a nephew, Richard, on October 13, 1648. This 140-acre gratuity consisted of several other plantations bought from the "persons hereafter following (vizt) of Ambrose Hancocke forty acres, of Henry Austine ten acres of a certain parcell of land commonly called Cold Harbour containing fifteen acres of land, and of Thomas Smyth and his present partner George Foster twenty acres."[21]

Big planters frequently dispossessed small and middling householders of the land in ways that left the latter bitter. One colonist observed that "the richer sort by giving credit to their profuse and sometimes necessitous neighbours on severe terms, insensibly in a few years wormed out the greatest part of the small proprietors."[22] In 1648 the church vestries were given the power to attach and appraise any of the "lands and housing of any person, or persons that do, or shall stand indebted upon any of their parish levies." Of greater importance, if "the said persons shall not satisfy the said levies and arrears, in some merchantable commodities," then the churchwardens "shall and may make sale of such lands and housing." That this system was open to abuse is obvious. Occupying land desired by richer and more influential land magnates—especially since they controlled the vestries—smallholders were subject to any number of exactions. In 1656 the bill was repealed, because it "hath been taken into serious considerations, how burthernsome, grievous and intollerable, it is to the inhabitants of this island, to pay assessments and levies which made, laid, and imposed upon them, by the several vestries of the parishes of the island, by a power granted them by a former statute, whereby one's estates have been exhausted and taken from them, to make payments of the said assessments and levies."[23]

Tenants also found their position undermined by the coming of sugar. The cost of rents rose, and planters were frequently unwilling to renew a contract because they saw greater profits to be had in sugar cultivation. Humphrey Waterman reported that his 800-acre plantation "had at one time 40 dividends [i.e., tenancies]" each inhabited.[24]

Most Christian bondsmen who gained their freedom in Barbados joined a growing class of landless freemen, who found it difficult to find employment in the island's slave plantation economy. Some became overseers and bookkeepers on large plantations, but planters were generally unwilling to hire freemen whom they thought were more expensive to employ than servants and slaves.[25] A small group of former servants were able to purchase or rent tiny tracts of land (one or two acres) on which they could practice little more than subsistence farming.

Adding to the trouble of the poor, the sugar revolution brought a dramatic rise in prices in the colony. Richard Vines complained in 1647 that "men were so intent upon planting sugar that they had rather buy food at a very dear rate than produce it by labour, so infinite is the profit of sugar works after once accomplished."[26] In 1651, an unknown writer, probably Gile Sylvester, a plantation owner, in an attempt to inform his family in England of the attractive returns to be had for European-manufactured goods in the colony, stated that "the Dutch sell their commodities after the rate of a penny for a pound of sugar. White or black hats yield here 120 lb of sugar, and 140 lb, and some 160 lb; pins at great rate and much desired; a man may have for them what he desired."[27]

Unable to find work and confronted with high prices, James Parker reported that the "common people" on the island "that have but mean estates, are very mean in respect of provisions, little flesh if any no bread but casader, a bread I approve not of; though it's true the rich live high."[28] Rather than exist in poverty, many colonists decided to leave Barbados for other settlements in the Americas. Capt. William Jackson, in 1645, was able to recruit 500 Barbadians to join him in a buccaneering adventure to the Spanish Main.[29] Capt. Robert Marsham enlisted another several hundred disillusioned colonists to populate the earl of Warwick's colony at Trinidad.[30] Around 1647, 1,200 islanders departed for New England.[31] While there was significant migration from Barbados by the late 1640s, it should not blind us to the fact that many stayed and suffered.[32]

Indentured servants were not the only ones who saw their dreams fade under the bright rays of the Barbadian sun. Also drawn to the island by the lure of profit was a middling class of planters. While they too did not have the capital to commit themselves to sugar immediately, they were in a position to pay for their own passage to the colony, with sufficient reserves to buy one or two servants and a modest parcel of land. They calculated that they could overcome the financial hurdles to sugar cultivation by what came to be known as the "Barbados custom."[33] That is, they first attempted to grow a staple that was cheaper to produce, such as indigo, ginger, or cotton, until they accumulated enough capital for the more costly venture into sugar. Richards Vines, a settler from Piscatagua, Maine, provides a vivid illustration of how the "Barbados custom" operated. In 1647 he migrated to the colony, where he purchased two adjoining plantations of fifty acres. He planted sixteen acres of his new acquisition in cotton. With the proceeds from the sale of his cotton, Vine told friends, "the next year I intend for sugar, at present I cannot."[34]

Even if he managed to put aside enough to finance the planting of a sugar crop, small and middling planters found sugarcanes a very risky staple. It had a long growing season, between fourteen to eighteen months; during this period, the plants could be destroyed by drought, hurricanes, rats, fire, and other natural disasters. Men of modest means could rarely afford to go so long without some returns on their labor; nor could they escape ruin if their crop failed.

It was mainly large tobacco planters, a well-to-do class of migrants who arrived on the island in the 1640s, and absentee investors who prospered by the sugar revolution. Big tobacco planters were in the best position to make the switch to sugar. Since they staked out their plantations when land was relatively cheap, they were not adversely affected by the escalation of land prices. Indeed, the rising cost of land worked to their good, since it increased the value of their holdings. In these boon conditions, some subdivided their property and reaped impressive profits by selling and renting it to the wave of new settlers that inundated the island.[35] While their farms were generally short of workers, big tobacco planters owned enough indentured servants to begin planting canes. Once they sold their crops, they brought in additional laborers and thus were able to enlarge the acreage they could cultivate. Owing to the size of their estates, they did not have to risk all in sugar, but if they were so inclined their contact with financial interests in England provided them with credit while their canes matured. The advantages that large planters enjoyed is perhaps best mirrored in the fact that, of the 175 largest sugar planters in Barbados in 1680, 40 percent of their families established themselves as members of the plantocracy during the tobacco era.[36]

The profits of sugar cultivation also attracted to the island a number of settlers who had the financial means to purchase estates that had already been transformed into a sugar works. Most conspicuous in this group was a class of prominent Royalist émigrés who, after the defeat of king and cause, fled England for the less oppressive atmosphere of Barbados. Still able to obtain credit in the mother country, they bought their way into the island's plantocracy.[37] Thomas Modyford is representative of these men. Modyford, a cousin of General Monck, at the raising of the king's standard received a commission to be a colonel in the Royal Army. Typical of most Royalist émigrés in Barbados, he was no blind supporter of monarchy. Agnes Whitson correctly characterizes his political career by saying that he "never thought it necessary to sacrifice himself or his family for an idea."[38] Captured at Exeter, where some Royalists charged that his duplicity had led to the fall of the city, he volunteered to leave England for Barbados. Not long after he reached the island, he purchased half of William Hilliard's sugar estate for £7,000 and became an influential figure in the island's politics.[39]

There were also many absentee investors who had an interest in a Barbadian sugar plantation. Prominent among London merchants were Maurice Thompson, William Penoyre, Martin Noel, and John Vincent.[40] Thomas Kendal, a grocer of London, purchased a third of William Hilliard's plantation.[41] Sir Anthony Ashley Cooper, later Lord Shaftesbury, owned a 200-acre sugar estate in St. George.[42] Not all absentee investors resided in England. John Parris, a Boston merchant, owned three plantations, land in Bridgetown, and a stone house at Read's Bay.[43]

In the 1640s, Barbadians could tell few rags-to-riches stories about the transition to sugar cultivation on the island. Since, for the most part, only men of substance could afford to run a sugar plantation, the sugar revolution continued the trend established in the tobacco era in which most of the island's wealth and productive capacity was controlled by a few planters.

Table 4.2 contains a summary of landownership by household in

TABLE 4.2
Land Distribution by Household
(1680s)

Parish	*Acres*	*Households*	*Households with land*	*Households without land*
St. Joseph	4,940	151	65% (98)	35% (53)
St. Thomas	7,713	275	68% (187)	32% (88)
St. John	7,637	165	67% (110)	33% (55)
St. Peter	6,803	577	25% (148)	75% (429)
St. George	9,569	200	54% (108)	46% (92)
St. Lucy	6,838	462	80% (371)	20% (91)
St. Andrew	5,408	172	47% (80)	53% (92)
St. James	6,820	204	79% (162)	21% (42)
St. Philip	13,151	510	79% (401)	21% (109)
Christ Church	14,461	425	91% (387)	9% (38)
Total for ten parishes 15,914	83,340	3,141	65% (2,052)	35% (1,089)

Sources: The censuses can be found in P.R.O., C.O. 1/44/142–379; Sloane Ms. (British Library) 2441/1–22.

the colony based on the censuses of 1680 and 1684. It gives a clear picture of the extent to which the switch to sugar perpetuated past inequities. Taking the ten parishes surveyed as a whole, 35 percent of all households (approximately 1 in 3) had no land. In the parish of St. Peter, 74 percent of all households were landless. Only 47 percent of the households in St. Andrew were in possession of land; 53 percent were not. Even in the parishes where there was a high percentage of households with land the figures are deceptive. In Christ Church, for example, 91 of the households held land, but 57 percent of them owned 10 acres or less, while 6 percent of the households possessed 60 percent of all parish land. In St. Lucy, to cite another instance, 80 percent of the households had land; 66 percent of those with a freehold, however, had no more than 10 acres. In the ten parishes, 47 percent of all households with land owned 10 acres or less, whereas 6 percent held 61 percent of the island's land. By combining households without land and those with a maximum of 10 acres, it is discovered that 61 percent of all households in Barbados had no more than 10 acres. This latter percentage would tend to suggest that the average white family

TABLE 4.2 (continued)

Households with 10 acres or less among all households		*Households with 10 acres or less among households with land*	*Households with 100 acres or more among all households*	*Acres held by households with 100 acres or more*
31% (47)	6% (296)	48% (47)	6% (13)	64% (3,180)
28% (77)	6% (459)	41% (77)	8% (24)	58% (4,438)
20% (34)	3% (213)	31% (34)	9% (24)	76% (5,797)
7% (43)	5% (353)	29% (43)	4% (23)	53% (3,621)
14% (28)	2% (227)	26% (28)	9% (31)	78% (7,484)
52% (240)	23% (1,583)	66% (240)	2% (10)	32% (2,187)
10% (18)	3% (136)	23% (18)	8% (22)	69% (3,724)
40% (82)	7% (472)	51% (82)	8% (23)	67% (4,581)
30% (153)	7% (945)	38% (153)	5% (30)	49% (6,465)
52% (220)	9% (1,351)	57% (220)	6% (31)	60% (8,657)
26% (972)	7% (6,035)	47% (972)	6% (231)	61% (50,134)

in Barbados could not expect to produce much beyond their basic needs on the lands they possessed.[44]

The pervasive use of African bondsmen on the island meant that there was a small market for free labor in the colony's economy, which contributed to the economic difficulties that Barbadian landless freemen and small planters experienced. By the middle of the seventeenth century, the consensus among planters was that slaves were cheaper to employ than white freemen, and as a result planters were reluctant to hire free laborers. It was not simply that they were convinced that black bondsmen made better field hands; they also saw economic advantages to employing slaves as artisans, so that even white craftsmen found themselves undercut by slave labor. The transition to sugar cultivation was largely responsible for the massive importation of African workers in the 1640s. It created a demand for labor far beyond the ability of England to satisfy.[45] Unfortunately, Africa had people for sale.

The history of the Atlantic slave trade began with the arrival of Portuguese explorer-merchants along the African coast in the mid-fifteenth century. The Portuguese did not come to Africa with a specific desire to purchase slaves. Their goals were more diverse in nature: to find a quick way to the Indies; to make contact with the mystical Prester John and his Christian kingdom; to establish a lucrative trade with local inhabitants; and, in the best of all possible worlds, to incorporate the people they encountered into a Portuguese overseas empire. In a short time they were trading for beeswax, hides, and ivory in Senegambia, and in the area that stretches from Assini in the west to the river Volta in the east (the Gold Coast) they found a wealth of gold. Also quick to develop was a market for slaves.[46]

Initially, as the Portuguese explored their way down the West African coastline, they would occasionally kidnap a few individuals from one of the local communities and carry them home—more as curiosities than as laborers. In the metropolis, however, a market for these slaves soon emerged, stimulating slave-trading activities in Africa. The gold trade further encouraged the Portuguese to buy bondsmen, since African merchants desired slaves in exchange for their precious metals. Giving additional impetus to Portuguese efforts to recruit bondsmen was the beginnings of sugar cultivation on the islands off the African coast (Cape Verdes Islands, Madeira, and São Tomé). It brought a demand for workers that Portuguese planters tried to meet by importing blacks from the African continent. Last, and perhaps most important, the birth of the Atlantic slave trade can be attributed to the Iberian penetration into the New World and the subsequent search for workers that it occasioned.[47]

As the demand for blacks increased, more systematic ways for raising cargoes of slaves were found. Frequently, the Portuguese would supply guns to one African community to help it defeat a bitter, historical foe. In return for their assistance, the Portuguese took as slaves any persons captured during the fighting. The price that merchants were willing to pay for bondsmen was a powerful incentive for individuals within African societies to find persons who might be sold to Europeans. Political factions found it a profitable and convenient way to rid themselves of habitual enemies. Greed induced others to sell kinsmen and rulers to barter away their subjects. Outsiders were frequently sold as slaves. Here the numerous African ethnic groups that inhabited West and Western Central Africa created an inexhaustible source of outsiders. Indeed, many who were taken as slaves in Africa were, in one form or another, victims of interethnic feuding. One ethnic group would sell persons of another community, and members of the former would in turn be sold by their neighbors. Through these various means Africans were drawn from their native land.[48]

By the third decade of the sixteenth century, northern European nations began to challenge the Iberian division of the New World and Africa. English merchants began appearing on the African continent in the 1530s, culminating in John Hawkins's efforts to intrude on the slave trade to the Hispanic colonies in the Americas. After the failure of Hawkins's venture, the English showed little commercial interest in Africa for the remainder of the century.[49]

It was not until the founding of the Guinea Company in 1618 that English attention was once again drawn to Africa. While the company would have been happy to establish a lucrative trade in any commodity, some of its prominent members were also involved in the colonization of the Americas, and they may have hoped to open up a commerce in slaves in the English Americas. Rumor had it that the earl of Warwick, a founding partner in the Guinea Company and one of the principals of the Virginia Company, intended to transport Africans to work in Virginia. If that was his plan, it came to nothing.[50] Perhaps the conflict that erupted in the Virginia Company, which would lead to the demise of that body, was related to the failure of his scheme.

English efforts to colonize the Americas were generally hampered by a shortage of labor, and with the Iberian slave colonies before them, it took a remarkably long time for Englishmen to experiment with slave labor. Some slaves were being used in Bermuda by 1617, but they were a minority in the work force.[51] Virginia did not import African laborers until 1619 (interestingly enough, Capt. Daniel Elfrith, who transported the first blacks to the colony, was an employee of Warwick at the time).[52] Yet it was not until the closing decades of the seventeenth century that the conversion to slave labor began in Virginia.[53]

In the Caribbean, the practice of using blacks occurred more quickly. In 1626, Maurice Thompson, a London merchant, carried sixty Negroes to St. Christopher.[54] In 1632 the Providence Island Company—of which the earl of Warwick was a member—decided to replace white servants with black slaves on Association Island, and in 1633 Negroes were introduced into its colony on Providence Island. By 1637 there were almost as many black slaves as free whites on the latter island.[55]

In Barbados slaves could be found in the colony from its first settlement, but because they cost more and were relatively harder to obtain than servants in the 1630s, planters tended to buy Christian bondsmen. The move to cultivate sugar dramatically affected the island's labor market: it unleashed a demand for labor that the trafficking in servants could not meet, provided planters with the financial ability to purchase slaves, and drew slave traders to the Barbadian coast. "Here are come lately about five hundred Negroes and more daily expected," one planter reported in 1642.[56] By 1645 George Downing could write that Barbadian householders bought that year no less than "a thousand Negroes, and the more they buy the better able they are to buy; for in a year and a half they will earn (with God's blessing) as much as they cost." A man hoping to settle on the island, Downing advised, reflecting the growing preference for black slaves over English bondsmen, had to procure servants, "which if you get them out of England, for six, or eight, or nine years time, only paying their passages or at most some small sum above it would be very well." This would allow a planter to start a plantation and in a short time with "good husbandry to procure Negroes (the life of this place)" out of the increase of his plantation.[57]

The Guinea Company was not up to the challenge of supplying slaves to the West Indian colonies. By 1628 the company had fallen into the hands of Nicholas Crispe, after it was abandoned by its original members. Crispe was not a force in the colonization movement. His interest in Africa seems to have been primarily with obtaining gold from the Gold Coast and redwood from Sierra Leone. With the company doing little to service the colonies with bondsmen, the growing demand for slaves in the English Caribbean brought interlopers into the African trade. Chief among these was Maurice Thompson. Through his connections, Thompson was able to gain admittance into the company, where he was influential in forging trading relations with the West Indian colonies.[58] By the 1640s the Guinea Company was becoming deeply involved in the transportation of slaves to the English Caribbean. In 1642, Nicholas Crispe and John Wood, the latter also a member of the company, sent to Barbados in the *Star of London* a cargo of African slaves that sold very quickly.[59] The earl of Carlisle in the mid-1640s was advised that he could have 200 slaves delivered to him in Barbados "by some London merchants." Carlisle was further told that English merchants had been participants in such transactions for blacks in the past.[60]

Though the company maintained a presence on the African coast for much of the 1640s and 1650s, and was jealous of its privileges, it was unable to prevent other Englishmen from trespassing on its monopoly. John Wadlow, a Bristol merchant, landed in Barbados in 1644 with a shipload of slaves for sale. Several Barbadian planters (James Drax, for instance) financed slaving expeditions to Africa.[61] There was no single or predominant place that interlopers frequented. Their ships appeared on the Gold Coast, the Niger Delta region, and the Congo.

In spite of the attempts to provide bondsmen for the colonies, neither the interlopers nor the Guinea Company was able to fully meet the needs of planters in Barbados. Thus, Dutch slavers found business opportunities on the island. In the first half of the seventeenth century, the Dutch had been more successful in their campaign to break the Portuguese monopoly of African trade. Starting in 1611, they built a series of forts along the African coast. By 1637 they captured Elmira Castle from the Portuguese, and in 1642 they effectively put an end to the Portuguese presence on the Gold Coast by taking their last remaining fort at Axim.[62] Consequently, when the English West Indian colonies began to employ slave labor on a large scale in the late 1630s, the Dutch were in a position to supply them with black bondsmen.

Between Dutch and English slave traders, approximately 135,000 Africans were carried to Barbados in the seventeenth century, at a rate of 2,400 per year.[63] The strength of the Dutch on the African coast would seem to suggest that they might have played a rather large role in the transporting of slaves to Barbados, but whatever advantage they enjoyed in the 1640s and 1650s, by the 1660s the English government successfully denied them access to the Barbadian slave market, making it an English preserve. Well over 80 percent of the Africans who reached Barbados as slaves in the seventeenth century arrived on English ships.

Between 1638 and 1650, Barbados was transformed from a depressed tobacco plantation system to a booming sugar economy. This new economic order supported a population estimated in 1652 at 18,000 whites and 20,000 blacks.[64] The main beneficiaries of the sugar revolution were a small coterie of large plantation owners. Owing to the monopolization of land and competition from slave labor, most free settlers found it difficult to reap part of the prosperity that sugar brought to the island. This led to a pattern of out-migration from the colony that would have serious implications for Barbados's political development. For those who remained, both white and black, the island's slave plantation economy prompted radical changes in their behavior and social outlook.

CHAPTER 5

The Plantation Household

The trickle of black slaves who came to Barbados in the 1630s posed no particular problems for planters. Most slaves lived in households in which they were the only person of African descent. They ate, worked, and slept with servants, and there was little effort to separate them physically from Christians beyond their exclusion from the church. Under these circumstances, they acculturated very quickly to plantation life, and planters had little trouble ordering them in their household. The mass introduction of slaves in the 1640s was quite another matter. Combined with the rapid expansion of the white population, it resulted in a cultural collision of monumental proportions. Groups of alien people, African and European, without historical ties or associations, sharing no common customs or values, were suddenly thrown into intimate contact on the island. The potential for misunderstanding and conflict was high.

While the degree of diversity should not be exaggerated, in the main, Africans and Europeans lived in cosmological worlds whose meaningful symbols, images, and codes of conduct differed substantially.[1] Linguistic barriers alone hampered social interactions. Yet, in a remarkably short time, through intimate contact and imposition, a shared system of norms and values that meshed with the realities of the burgeoning plantation economy was created. In the colony, many old world taboos lost their meaning or were transformed into public virtues, as individuals adapted to new notions of self and community. In this new order, Africans did not become Englishmen, nor did Englishmen become Africans. Each retained portions of their traditional heritage, but the demands of the plantation system encouraged creolization (i.e., the formation of new cultural identities and behavioral patterns). By the end of the seventeenth century, there were decided changes in the ethnic consciousness and conduct of Africans and Europeans.

Central to the creolization process was the plantation household. It was within this institution that Africans and Europeans forged many of the relationships that were to ultimately lead to the emergence of new social patterns.

The plantation household was the single most important multiracial institution in the Americas before the twentieth century. While its multiracial character, and its bipartite kinship and familial segmentation, would have been enough to differentiate it from Old World household arrangements, it also differed functionally from traditional households in Africa and Europe. Its uniqueness would seem to support a growing consensus that plantation societies did not resemble capitalistic or feudalistic societies but were a distinct social system that fostered forces peculiar to its internal structure.[2] If the history of Barbados is any example, the appearance of the plantation household can be taken as a manifestation of the profound social and cultural changes that Europeans and Africans experienced within a plantation economy.

Much of the daily existence of slaves was spent on their master's plantation. Their homes were thatched huts with dirt floors built in close proximity to the more imposing structure that their owners occupied.[3] This residential pattern caused the plantation to resemble a village. Father Antoine Biet wrote in 1654 that "most plantations in the country are like as many villages whose size varies according to the number of slaves each plantation has."[4] While Thomas Trapham's description of a Jamaican plantation in 1678 is far more detailed than Biet's, it conveyed the same impression that sugar estates were essentially villages. "The stranger," Trapham observed, "is apt to ask what village it is (for every completed sugar work is no less), the various and many buildings bespeaking as much at first sight, for besides the . . . Mansion House with its offices, the works, such as the well contrived mill, the spacious boiling house, the large receptive curing houses, still house, commodious stables for the grinding cattle, lodging for the overseer and white servants, working shops for the necessary smiths, others for the framing carpenters and coopers: to all which when we add the streets of the Negroes houses, no one will question to call such a completed sugar work a small town or village, as well as for the number of inhabitants as buildings."[5]

Biet and Trapham were basically struck by how much the spatial arrangement of a sugar estate resembled a village. Had they looked closely at its social infrastructure, they would have been all the more convinced that sugar plantations were village communities, not simply farms and factories organized for the production of sugar. As previously mentioned, for their residents, in particular those in servitude, it also functioned in some fashion as a court, a school, a hospital, a workplace, and a home.[6] The plantation household was the sum total of this complex network of interconnected social and economic relationships. It was dominated by a planter-patriarch, who used the various components of the household to govern the lives of his servants and slaves. Since the state in Barbados tended not to interfere in the general management of plantations, particularly in those areas related to the governance and treatment of bondsmen, the planter's ability to regulate the activities of servants and slaves was quite extensive. He dispensed justice and decided who acquired what skills and how food and clothing allowances were distributed on his estate. Dependent laborers, however, were not pieces of clay that planters could mold into any shape they pleased. In essence, the structure and function of the plantation household was a compromise between the master's ambition to produce sugar with servile labor and the struggle of those bondsmen to preserve a human existence. Of central importance to this study is the fact that inequalities in the plantation household promoted tensions between Africans and Europeans that spilled over into the larger society, where they helped to give life to a racially conscious, brutally repressive political regime. It was a government of minority rule, which quickly became dependent on the imperial system for its maintenance.

It took two to three weeks to sail from the African coast to Barbados, but for most slaves it must have seemed like an eternity. For much of the trip, Africans were confined to the ship's hole, where they were forced to sleep in their own excrement, chained to men whose anguished faces mirrored their own anxiety. The portholes would be opened in fair weather, and a steady stream of fresh air would seep in, bringing some abatement to the haunting, foul stench that lingered everywhere. And then there was death. By the time a slaving vessel arrived in the New World, pestilence and disease were usually running rampant on its deck. It was not uncommon for a slaver to lose between 20 and 30 percent of its cargo—all in a matter of weeks.[7]

The survivors of this experience were in no position to resist the forces of change that engulfed them in Barbados. As noted earlier, taken together, they had no sense of themselves as a black or African people, and no past history of coordinated action. In their minds they were Ibos, Ijo, Mbundu, or one of the other African ethnic groups represented on the island. This heterogeneity severely retarded the Africans' ability to respond collectively to their enslavement.[8]

The diversity of Africans may have made their incorporation into the plantation system easier, but holding them in bondage was still no simple feat. In fact, though planters purchased Negroes, called them members of their household, and made them work in their fields, in English law there were no provisions that defined the status of slaves.[9] Before Barbadians passed an act in 1636, declaring "Negroes and Indians, that came here to be sold, should serve for life, unless a contract was before made to the contrary," no statute answered even the most basic questions regarding slavery.[10]

In the absence of legal guidance, slaveholders worked from the assumption that through purchase Africans became their property. As their possession they could use these blacks as they liked, and the children of slaves were also theirs to keep and control. The idea was to support Negroes at a standard that would permit them to carry out their employment. This allowed the owner to appropriate for his personal use all that did not go to bondsmen for their maintenance. The words of Morgan Godwin, a sharp critic of Barbadian planters, are enlightening in this matter. Godwin observed that planters gave their slaves provisions, not as "their right or due, but as conducive to their masters' convenience and profit." "They considered it only in order to the enabling of their people to undergo their labour, without which themselves cannot get riches, and great estates."[11] Godwin was no friend of slaveholders, as the tone of his remarks suggests, but this should not blind us to the basic point he was making. Among Barbadian planters it was thought that a successful slaveowner weighed the balance between what he gave slaves and what he received from them.

By the very nature of slavery, all blacks, even the most enterprising, were confined to poverty and had little hope of escape. In the island's plantation economy, they were a permanent underclass, completely under the authority of their owners. However, compassionately used, the discretionary powers that masters had over their slaves was staggering. It was their right to decide the occupation of their bondsmen. Planters determined how long a slave labored during the day, how long he rested, how much he ate, and what he wore. The integrity of slave families in their household depended on their will. Punishment, reward, and at times life or death were the masters' to give or take.

Until the late 1660s planters argued that the paganism of Africans made it proper for them to be held in bondage. Well before the settlement of the colony, it was commonly thought, though never scrupulously followed, that one Christian could not enslave another. All those beyond the pale of Christianity were fair game. Following this reasoning, Englishmen justified the enslavement of Africans in Barbados.[12]

In classical African terms, planters would have been viewed as witches and sorcerers, who had the capacity to cause harm and promote their own interest, either directly or indirectly, by harnessing the power of one of the spirits. Thus it is not surprising that a slave in Barbados thought that the "devil was in the Englishman, that he makes everything work; he makes the Negro work, the horse work, the ass work, the wood work, the water work, and the wind work."[13]

While planters did not stop slaves from praying to their gods, the demographic composition and the social organization of the island's plantation system worked to inhibit the evolution of a mass religious movement among Africans (similar to those that developed in their native lands) that might have challenged the authority of the plantocracy. Plantation life left blacks with frustrations that made them a socially dangerous and unknown quantity. It did not take a servile revolt to convince Anglo-Barbadians of the rather large possibility that their slaves might rebel. English political culture inculcated the notion that poor men were prone to crime and rebellion, and who could have been more impoverished than their slaves? Planter fears and black dissatisfaction were written in the architecture of plantation society as slaveowners constructed their homes with easy access to water so that if "an uproar or commotion occurred on the island they could use it to drink as well as to throw down upon the naked bodies of the Negroes, scalding hot; which is as good a defense against their underminings, as any other weapons."[14]

The diametrically opposite aspirations of masters and slaves in the plantation household produced contradictions that threatened to tear the institution asunder.[15] If their estates were to operate as profitable enterprises through the use of African bondsmen, planters not only had to persuade or coerce an alien slave population to work productively; they also had to simultaneously arrest the tendency toward disequilibrium that existed within their household. Indeed, Barbadian patriarchs used profit and security as the principal yardsticks by which they distinguished between permissible and illicit behavior among their household slaves. To be sure, these objectives were not the sole determinants of all acceptable and unacceptable conduct; but, speaking in the aggregate, had planters permitted the behavior of slaves to drift away from these priorities, it is likely that the disposition for instability, as represented in the dreams of the black population, would soon have challenged the existence of the plantation household. It was through their broad discretionary powers that masters sought to keep the antithetical forces that were generated by the structure of their households from manifesting themselves in a social upheaval. They punished all actions that they perceived as a threat to profit or security, and they reinforced conduct that worked to their interest through a sophisticated regime of rewards.

Col. Henry Drax, the second-wealthiest planter in Barbados, set down his philosophy regarding the punishment of slaves in a list of instructions to his head overseer. "You must never punish either to satisfy your own anger or passion," Colonel Drax explained. The end of punishment was either to "reclaim the malefactor or to terrify others from committing the like fault."[16] The behavior that warranted chastisement was not limited to recalcitrancy or a failure to work hard enough. An error at an assigned task could be the cause of a beating. Drax suggested, for example, that the head boilers must discharge their duties carefully, and upon their "neglect to be severely punished." Recognizing that bondsmen were often reduced to pilfering plantation goods to supplement their meager allowances, Drax thought that if slaves stole "for the belly, it is the more excusable." "But if at any time they are taken stealing, sugar molasses or rum, which is our money and the final product of all our endeavours," he strictly ordered, "they must be severely handled being no punishment too terrible on such an occasion as doth not deprive the party of either life or limb."[17]

On those occasions when it was necessary to discipline a bondsman, Drax recommended that the correction be severe and that it be administered at the time one took notice of misconduct, "many of them [Negroes] being of the humor for avoiding punishment when threatened to hang themselves."[18] The propensity for slaves to commit suicide was related to the torture and mutilation they were sometimes subjected to. Father Antoine Biet described how one planter had a slave whipped by other Negroes repeatedly for seven days. On the eighth day he had one of the slave's ears cut off and roasted. The black was then made to eat his ear. Although Biet agreed that "one must keep these kinds of people obedient," the gruesome spectacle caused him to write that "it is inhuman to treat them with so much harshness."[19] Barbadian slaveholders believed, however, that they had to "exceed the limits of moderation" in the punishment of blacks so as to "intimidate and impress fear and dutifulness upon them."[20]

It is hard to assess how frequently violence was employed as a means of social control on Barbadian plantations; but in all probability it was not a daily occurrence. Capricious terror would have been self-defeating, persuading slaves that they would be tormented whether they worked or not. At the very least, the slaveowner's investment in his bondsmen put some curb on abuse. The fact of the matter is that constant resort to violence was not necessary; planters had other means by which to regulate their slaves' behavior.

The quality of a slave's life (the type of job he performed on the plantation, the clothes he wore, the well-being of his family) was in great measure determined by his master, and the bondsman who lost his owner's favor was destined for hard times. Planters used their control over the necessities of life so as to construct a system of rewards that reinforced behavior that increased the profitability and security of their household.

On large plantations, they employed it to create an administrative hierarchy through which they governed their households. Just below the master and his family in this structure were indentured servants. In the 1630s, when few owned slaves and most work was done by servants, there was little effort to discriminate against slaves beyond their exclusion from the church. The mass importation of Africans in the 1640s radically altered the demographic composition of the plantation household, leaving planters dangerously outnumbered by their slaves; in some instances, the figure reached as high as 13 to 1. It also increased social tension that could easily have been manifested in an insurrection of slaves. Adjusting to these new problems, planters elevated servants to a privileged class in their household.

The very fact that servants were Christians in a society where only non-Christians could be slaves immediately set them apart from Africans. Perhaps the best way to understand the implications of being a Christian in seventeenth-century Barbados is to see how it translated into everyday social relations on the plantation. By a law passed in the early 1640s all masters were required to bring their servants together every Sunday for morning and evening prayers.[21] On the Drax-Hall and Irish-Hope plantations it was common on Sundays for "all the whites in the family [i.e., household]" to be "called in to hear morning and evening prayers."[22] Servants who failed to attend forfeited their allowance of food for the week. One can imagine the effect it must have had on servants and slaves as the former gathered with their masters in religious services. It established a line of division between servants and slaves. To further solidify the stratification, playing on African religious beliefs, planters made Christianity appear to be a secret society that endowed all its adherents with great knowledge and privileges. To instill this notion in the minds of slaves, blacks who worked in their master's house were taught to leave the room whenever Christians said their prayers, "as if there were some secret charm, or power of doing mischief in prayers," confirming African suspicions that Europeans were witches and sorcerers.[23] Confronted with such ideas and practices, those Africans who lived in the colony long enough to learn that their religious beliefs served as the legitimation for their enslavement began to request that they be allowed to convert to their master's faith. Richard Ligon tells of a slave named Sambo who desired "that he might be made a Christian for he thought to be a Christian was to be endued with all those knowledges he wanted."[24] Sugar producers, however, were violently opposed to the Christianization of their slaves, which they thought would jeopardize the institution of slavery, because "being once a Christian, he could no more account him a slave."[25] Whenever the subject of evangelizing among black bondsmen was broached, it was commonly retorted, "what, shall they be like us?"[26] In the earlier years of the plantation system in Barbados, religion helped to distinguish superior from inferior persons. To describe a person as a heathen came in popular parlance to mean a slave and a black with all the attending social implications.

As Christians, servants formed a privileged body in the plantation household. On estates where the conversion to slave labor was completed, Christian bondsmen were no longer employed as field hands. They were now the overseers, bookkeepers, headcurers (responsible for refining the juice of the sugarcane into sugar)—in short, they became members of the skilled and managerial class in the planter's household. In these capacities, it became their responsibilities to see that blacks performed their assigned tasks, making them a symbol of the master's authority.

Although servants as a class were privileged in comparison to slaves, they were not all treated alike. Depending upon the skill of the servant, and the importance of his occupational role on the estate, he could be given extra allowances, even small wages by which he might put a little aside for the day when he acquired his freedom.[27] Certain Christian bondsmen ate at the master's table to demonstrate their closeness to the patriarch, which was of great symbolic significance. Following his own practices, Drax instructed his principal overseer, whom he was leaving in charge of his estate while he went to England, that since the headcurer would "have great trust commanded to his charge," he should be "above conversing with the ordinary sorts of servants for which reason I would have him eat at your own table."[28] Drax further ordered that the plantation doctor and bookkeepers also be given their meals with the head overseer.[29] In this way, they formed a kind of "kitchen cabinet" to advise the overseer on the daily operation of the estate.

Although as a rule no slave was permitted to dine at his master's table, planters did bring blacks into the administrative structure of their household. They were employed as drivers, artisans, and cartsmen. Blacks who held positions of trust on the plantation were allowed liberties that were denied other slaves. Macow, a slave of Thomas Modyford, was permitted to enter his master's house, "which none of the other Negroes used to do unless an officer, as he was."[30] When it came to selecting wives, planters gave elite slaves first choice, and "so in order as they are in place, every one of them knows his better, and gives him the precedence."[31] Blacks with extraordinary skills were allowed two or three wives.[32] Plantation owners also showed special favor to the families of trusted blacks. Henry Drax directed his overseer to keep Monkey Nouo, whom he described as "one excellent slave," in his position as head black driver. In addition to the normal dietary allowances that Drax provided his slaves, Nouo was to have ten pounds of fish or flesh a week, which he could dispose to his family as he saw fit. Nouo was also to have an extra allotment of clothing for himself.[33] In return for these indulgences, patriarchs expected that privileged slaves would be more trustworthy than the average field hand.

An incident that occurred on Thomas Modyford's plantation illustrates the extent to which some elite slaves identified with their master's interest, and how important they could be to the stability of a sugar estate. The year 1647 was a "time when victuals were scarce"; planters had to cut back on the rations that were given to bondsmen. Discontented with the reductions in their food allowance, a few of Modyford's Negroes plotted to set fire "to such part of the boiling house as they were sure would fire the rest, and so burn all, and yet seem ignorant of the fact, as a thing done by accident." Their scheme, however, was discovered by some of the planter's more trusted slaves, who reported the conspiracy to their master. Modyford ordered "condign punishment" for the rebellious blacks; the rest of his slaves were to be given a day of liberty, that is, a day of leisure, and a double proportion of provisions for three days. The slaves refused to accept their owner's indulgences. "Knowing well how much they loved their liberties and their meat," and "suspecting some discontent amongst them," Modyford "sent for three or four of the best of them," and asked these elite slaves why they had refused his favors. He was told that "it was not sullenness or slighting the gratuity their master bestowed on them, but they would not accept any thing as a recompense for doing that which became them in their duties to do." Furthermore, Modyford was informed that they would not have him think "it was hope of reward that made them accuse their fellow servants, but an act of justice, which themselves bound in duty to do." They did, however, tell Modyford that if it pleased him "at any time to bestow a voluntary boon upon them, be it never so sleight, they would willingly and thankfully accept it." This report, "in such language as they had," they delivered to their master. The principal spokesman was Sambo, "by whose example the others were led in the discovery of the plot and refusal of the gratuity." Sambo was one of Modyford's drivers, who constantly received indulgences from his owner, and in turning in the conspirators he further established himself as one of his master's faithful bondsmen. Indeed, the events led Ligon, who was staying on Modyford's plantation, to declare that "there are to be found amongst them [slaves] some who are as morally honest, as conscionable, as humble, as loving to their friends, and as loyal to their masters as any that live under the sun." But even Ligon suspected that if blacks, including trusted slaves, ever thought they had the "power or advantage," they would "commit some horrid massacre upon the Christians, thereby to enfranchise themselves, and become masters of the island."[34]

The average slave on a sugar estate was usually incorporated into one of the plantation's gang. The sources are silent on precisely when and how the gang system was established in Barbados.[35] In all likelihood, it began in the 1640s when planters, faced with an alien labor force that spoke no English, found it convenient to have them labor in large groups. Henry Drax was of the impression that the best way to "prevent idleness and to make Negroes do their work properly" was to organize them into gangs. How these gangs were constituted depended upon the work to be done. During the planting season, adult male slaves were divided into two gangs: "the ablest and best by themselves for holing and the stronger work, and the more ordinary Negroes in a gang for dunging [i.e., manuring the fields]." Women were likewise separated into two gangs according to their physical abilities. On established plantations, such as Drax-Hall, they prepared the land for cultivation by removing weeds and tilling the soil. Children were also divided into two gangs. Toddlers and infants were placed under the supervision of a woman who was too old for field duty. Little work was expected of these infants. Older boys and girls were recruited into a gang to do odds and ends around the plantation (i.e., weeding, feeding the livestock, and sometimes killing rodents that preyed on the cane stalks).[36]

The harvest season brought some changes in the gangs. Slaves were separated into those who worked in the mill, the boiling and curing houses, and the cane-cutting gangs. All the other Negroes (except toddlers) were placed in a "running gang" to fetch and store all sorts of fuel for the boiling house. When there was no work for the boiling house and the mill house blacks, they were sent to labor in the running gang.[37]

All the gangs, whether during the planting or the harvest season, were led by a black driver who kept a "list of the gang under his particular care that he may be able to give a particular account of everyone, whether sick or how employed."[38] Every fortnight, and sometimes more frequently, Drax had these lists brought to him so that he might know "where every Negro in the plantation was employed." This system, he wrote, was "most effectual to prevent lazy Negroes absenting [themselves] from their work."[39]

The structure and the priorities of the plantation household, which differed markedly from those that supported European and African familial relations, prompted new social patterns in Barbados. Gone for the slave were the daily rhythms of the African calendar (harvest times and ceremonial days) and the community of kins and neighbors who reinforced tribal life-styles. In their place were the crop times of the sugar estates and their master's imposing authority. This is not to paint a stark picture of the total deculturation of the African in Barbados. Where Old World customs and ideas impinged on the functioning of the plantation household, slaveowners worked to destroy them. For the same reason, those African beliefs or practices that helped to maintain the plantation were encouraged by planters. Thus, it would seem more accurate to say that Barbadian slaves still employed some Africans beliefs and customs, but the requirements of the plantation imposed new patterns of behavior on Africans, which in time, as will be seen later, left them with a much altered cosmology.[40]

Surrounded by an alien and potentially hostile black population, Englishmen also had their world rearranged by the pressures of a plantation economy. One planter, Ligon recalled, had 200 slaves in his household whom he had to feed daily, "and keep them in such order as there are no mutinies amongst them; and yet of several nations." Besides, "all of these are to be employed in their several abilities so as no one be idle."[41] To prevent the internal contradictions from destroying their household, and to ensure that it continued to show a healthy profit margin, planters not only sponsored social relations and ideas that prompted those ends; they also participated in these relationships, and their world was modified accordingly. A good example of the adaptations that Englishmen made in plantation society may be observed in the political system they forged in order to preserve the inequalities in their households.

Even with the broad discretionary powers that planters were able to wield on their plantations, a master who owned 50 or more slaves, to say nothing of those with 100 or 200, would have found it impossible to keep his bondsmen under control without assistance from the state. Nor would an individual planter have been able to do much in the face of a general conspiracy, with slaves from the various plantations acting in concert to gain their freedom. Thomas Moore must have been painfully aware of this fact when he appeared before the Barbadian Council in 1657 to plead for aid against the "divers rebellious and runaway Negroes" who were lurking around his plantation. The blacks, he told the Council, committed many violent acts and attempted to "assassinate people to their great terror."[42] If the inequities that were inherent in the plantation household were to survive, the state had to play an active role in their preservation. In working to secure the master/slave nexus, the objective of the government in Barbados drifted away from traditional English political values.

In the mother country, though sharp inequalities existed, there was the basic sense that the future of an independent England depended on economic prosperity and the survival of a large, interested population. Consequently, when the English crowd began to demonstrate against engrossers (those who buy in large quantities with the view of being able to secure a monopoly), forestallers (those who withhold goods from the public market in the hope of creating a scarcity), and the like, rather than suppress the mob without giving a thought to their complaints, the English government tended to treat the rioters with paternalistic indulgence. The combined effect of mob action and government tolerance was to produce a "moral economy" that inhibited innovativeness or acquisitiveness that might have left England with a population that was "driven out of all heart."[43] The crowds defended their material interest while the authorities looked away, understanding that the flourishing of a state consisted "chiefly in being strong against the invasion of enemies nor molested with civil war the people being wealthy."[44]

In Barbados the continued subjugation of slaves became as much a political imperative as the perpetuation of an interested population was in England. From the far too simple decision by one individual to enslave another came the practical political problems of how to maintain the relationship. Rules were demanded; otherwise, the volatile union could have exploded, and it was here where the state intervened. It used the plantation household as the first line of defense against social turbulence by charging masters with specific responsibilities regarding the behavior of their slaves. It directed that masters were not "to give their Negroes leave on Sabbath days, holidays, or at other times to go off their plantation." Exempted from this order were trusted slaves, "who were required to have a ticket signed by their owners." The pass had to state the time the slave left his master's house and when he was supposed to return. To keep blacks from hiding runaways from other estates, it was ordered that masters had to search their slaves' quarters twice a week, and every fortnight slaveholders were instructed to look for "clubs, wood swords, and other mischievous weapons."[45]

It was not simply a matter of keeping weapons out of the hands of slaves; the government decried as a public nuisance any ideas, or conveyers of such ideas, that caused slaves to ponder their right to freedom, as is evident in the harassment of Quakers in 1675. After the discovery of a slave insurrection in that year, a general apprehension seized the government about Quakers who were permitting bondsmen to attend their meetings. The fear was that blacks who were participating in these religious ceremonies would be taught in principles "whereby the safety of the island may be much hazarded."[46] When Governor Atkins brought the matter before the General Assembly for its consideration, he spurred that body into action by raising the question of "whether liberty be a doctrine of slaves."[47] The resounding answer was no, and penalties were instituted to prevent slave attendance at Quaker meetings.[48]

While the state did much by way of legislation to secure the master-slave relationship, it accomplished even more by what it did not do. In fact, the state tried to interfere as little as possible when it came to the subject of planter control over the bondsmen in their household. In effect, it legitimated the expansion of patriarchal authority that occurred in the formative years of settlement. There were no laws that specified the amount of food a slave should receive; there was nothing that limited the nature and quantity of punishment; nor was there any statute that established days of rest. In short, the colonial government regulated heads of household in their employment of slaves only insofar as public order was concerned.[49]

The perennial question before the island's government was how to retard the proclivity for instability that was present in the plantation household. The problem had broad implications for the island's political history, as the use of political power in Barbados was informed, not by English concerns for an interested population, but by the pervasive fear of a slave revolt. The state tried to exorcise the slightest hint of a servile insurrection out of the society with castrations, hangings, and deportations.[50] It constantly warned planters that overindulging blacks and inattentiveness were the parents of most rebellions, and to prevent the unthinkable it sanctioned social customs and promoted ideas that perpetuated the stratifications in the colony.[51]

In coping with the tensions and strains produced by the slave plantation system, planters built a government of minority rule that could not confidently look to the majority of local inhabitants for support in times of emergency. This fundamental fact was of central importance for the structure and stability of Anglo-Barbadian relations.

PART II

THE POLITICAL ECONOMY OF SLAVERY

CHAPTER 6

A Precarious Independence

From its founding in 1627 until it gained independence in 1966, Barbados was a dependent colony of England. For one brief moment in the seventeenth century this relationship faltered.[1] Between 1636 and 1646 a series of political events occurred that destroyed the proprietorship and that severely reduced England's effective control over the island. In the main, it is a story of how the plantocracy used the general dissatisfaction with proprietary government on the island and the outbreak of civil war in England to free the island from any specific governing force in the mother country. Related to these political maneuverings was the rise of sugar cultivation in the colony. Gazing upon the divisions in the metropolis, planters realized that by taking sides in the English disputes they would be inviting upheaval in Barbados at a time when the introduction of sugar was promising future prosperity. This fear of a contagious reaction led them to neutrality and an independent course in their dealings with England.

During this period, planters learned to appreciate the benefits of home rule; they welcomed the fact that legal disputes could be resolved in Barbados with no appeal to England, and they experienced the advantages of having local officials who were directly accountable to them.

The island prospered greatly by the free trade that it conducted, and sugar producers wanted to see it continued. By the time a unified central government reappeared in England, many colonists were openly advocating home rule. Even the most moderate planters wanted the island treated as if it were an English town or corporation with the right to elect its own local officials and to elect representatives to Parliament, but little was done to prepare the society to defend the freedom that it had accidentally acquired. In fact, the slave plantation system that emerged during this period of independence unleashed a set of social and political problems that ensured the effective reintroduction of colonial rule.

Hoping to put an end to the political instability that had become almost commonplace in Barbados in the late 1630s, the landlords appointed Philip Bell deputy governor of the colony. Bell was an experienced West Indian pioneer well accustomed to the problems of colonial administration. He had previously served as governor of Bermuda (1626–29), Providence Island (1631–36), and St. Lucia (1640–41).[2] As his connections with Bermuda and Providence Island would seem to suggest, Bell had close ties with powerful Puritans in England (particularly the earl of Warwick), but in leaving his post on the latter island he quarreled bitterly with his former employers, and thus in no sense can he be considered their agent in Barbados.

During his tenure as deputy governor, Bell guided the island through the turmoil of the sugar revolution and the destruction of proprietary government.

Simultaneous with the changes in government, Carlisle, with the approval of the trustees, required heads of households to take an "oath of fealty," swearing to be obedient tenants. Carlisle also issued a proclamation in which he tried to calm fears that the proprietors might dispossess the settlers just as the introduction of sugar was about to transform the island. He promised that "the proper estates in land of every man there legally interested shall remain to him and his heirs, in fee simple, notwithstanding any sequestration or misreports to the contrary."[3] The proprietors even showed a surprising willingness to negotiate with the colonists for the rents. Unaware that the collection of the rents had reverted back to the old system, the earl announced that he was leaning toward ratifying the original arrangement struck between Huncks and the Assembly, but he let it be known that he would entertain counterproposals.[4] Perhaps matters would have worked out better for the landlords had they shown a readiness to compromise with their tenants earlier; but with central authority about to disappear in England, Carlisle's probing ended in disaster for the proprietorship.

Householders quickly seized upon the invitation to bargain for the rents. In the fall of 1641, the Barbadian Assembly forwarded some proposals to the landlords. Despite their anxious anticipation, by January 1642 the islanders had not received any response to their propositions, even though the proprietors had been warned that the inhabitants "will pay nothing" till hearing from England.[5]

The delay was in part caused by Carlisle, who was again trying to outmaneuver the trustees for control of the colony. In January 1643 he convinced the king that the Barbadians had "all turned to the way of Parliament" and "were like to shake off their obedience to the crown." Believing this misinformation, Charles I acquiesced in Carlisle's decision to recall Bell and appoint Edward Skipwith, a courtier, in his stead.[6] Throughout Carlisle's machinations, Archibald Hay was absent from court. Upon his return, the trustee was able to assure Charles that the colony had not revolted against the crown and allied itself with Parliament, but the damage had already been done.[7]

Hoping to cut off reports about his intrigues to replace Bell, in March, Carlisle wrote the governor that he need not fear dismissal. It was true, the earl told Bell, that the king forced him to prepare a commission that deputed Skipwith governor of Barbados; but, in consideration of Bell's fidelity, and "out of more exact knowledge of Skipwith's insufficience," Charles allowed him to withdraw the grant.[8]

In the same letter Carlisle turned down the Assembly's 1641 offer on the rents, saying that it was "unreasonable and too strict." He hoped, however, that the rejection would "produce no difference twixt me and them when God brings us together and we may better understand one another."[9] Through some private courier the earl passed an intimation to Bell that "if he could draw them [the Assembly] to three pounds per acre again (as was formerly offered) that for his part he would not so far as lay in him dissent."[10]

Before Carlisle's letter reached Barbados, after two years of patient waiting for a reply, and no doubt disturbed by reports of his replacement, Governor Bell decided to submit to a petition from the inhabitants that he summon a general assembly. At the convening of the representatives, the governor surrendered his share of the proprietary rents, accepted an annual allotment of £1 per acre for his own support, and acceded to an act halting the payment of the rents to the proprietors.[11] By receiving the annual stipend from the General Assembly and consenting to the abrogation of the rents, Bell undermined proprietary authority in the colony and became a creature of the Assembly, upon whom he was dependent for his yearly salary. In effect, a palace coup took place; the proprietary government was overthrown, and the plantocracy exercised full power in Barbados.

Having lost his lucrative position by the interruption of the rents, Receiver Browne was infuriated. Bell, he insisted, in approving these acts, "waived and declined" the power granted to him by "vilifying and annihilating" proprietary authority, "instead of exacting obedience from this perverse generation." Even if the governor were dismissed, as Browne strongly urged, he surmised that it would "prove a matter of more than ordinary difficulty to rectify the distempers which timidity and cowardise have offered ever uncontrolled to mount to that height of insolence which long time and policy can hardly reform."[12]

Bell excused his collaboration with local planters by arguing that he and the whole country had waited with "great expectation some satisfactory answer" to the propositions. In this manner they continued until the colonists became tired, and "publicly and generally protested an unwillingness, and indeed resolution, to pay no more rents at all." This, coupled with the disorders at home and the belief that the power of the proprietors were revoked by Parliament, caused the inhabitants to grow to a "desire and pretence of a common liberty, and indeed an exemption from all authority and obedience." Since there existed no legitimate authority "for all the protection of law or justice, no means to keep or maintain any sessions, no ammunition either for exercise or defence, no fortification and such like," public safety, Bell contended, compelled him to call the General Assembly together in June 1643.[13]

Once seated, the General Assembly conducted itself as the sovereign entity invested with legal responsibility for governing Barbados. It declared, for example, that none of the said articles, statutes, laws and ordinances [of Barbados], shall at any time hereafter be repealed, or nullified, in part, or in whole, nor anything thereunto added, without the assent, consent, advice, and approvation of a like general assembly, consisting of the governor, council, and freeholders.[14]

Asserting their right to make and keep laws that none other than themselves, including the proprietors who went unmentioned in the act, could alter or repeal was a major step toward independence. An even bolder step was the planters' claim that they were "freeholders." The problem of land tenure had long haunted Barbadian householders. So, in two acts, the Assembly proclaimed that

all the inhabitants of this land, that were in quiet possession of any lands or tenements by virtue of any warrant from any former governor, or by conveyance or other act in law . . . are . . . adjudged and declared to have and to hold their lands of right, them to dispose of or alienate, or otherwise to descend, or be confirmed to their heirs forever.[15]

In these statutes, of dubious legality in English law, members of the plantocracy usurped the rights of the proprietors and established themselves as the sole legitimate rulers of the colony. Had there existed a strong central government in England able to command faithful adherence to the proprietary patent, it is unlikely that these statutes would have passed in the Assembly, and even less likely that they would have been enforced.

Upon assuming political control over the island, the plantocracy did not immediately terminate all negotiations with the proprietors. In June 1643, Bell presented Carlisle's belated counterproposal of three pounds of cotton or tobacco per acre as the rent. In September the Assembly announced that it would agree to this sum, if the proprietors would (1) use all unpaid rents, calculated at "near 60,000 pounds of tobacco or cotton," for public purposes, particularly the fortification of the colony; (2) pay the expenses of all planters who sat at Council meetings, which some thought would encourage the governor to call them into session more frequently; and (3) remove all of their supporters from the Council. This latter was to keep secret, Browne charged, "any private practices that may be concluded on in behalf of the country." Foremost, planters demanded that they have a "free and absolute confirmation" of all their lands and plantations.[16] Had the earl agreed to these terms, Barbadian householders would have become secure freeholders with legal control over the Council and the Assembly.

The colonial challenge to proprietary rule provoked a flurry of activity in the mother country as the earl of Warwick came to see this as an excellent opportunity to capture this flourishing colony from Carlisle. Thirteen days after a group of merchants and absentee planters complained to Parliament that Carlisle had much "hindered the plantation" and that the king was seeking to draw Barbadians to the "popish party," an ordinance was passed establishing a committee of foreign plantations to govern the English Americas. By this statute, the earl of Warwick was named governor in chief and lord high admiral of all American plantations. Under the auspices of Parliament, Warwick obtained the legal interest in Barbados that he had been unable to acquire in the 1630s.[17]

In an effort to cajole colonial acknowledgment of his right to rule, soon after taking office Warwick sent several orders to Barbados exempting the inhabitants from all taxes and common charges, "other than what should be necessary for the support of the government and defraying the public occasion of the island." Warwick also authorized Barbadians to choose their own governor, subject to his and the committee's approbation. In exchange, the earl expected the colonists to oppose any government not approved by him; nor should they allow anyone to "exact any taxation other than what is before mentioned."[18] By encouraging planters in their rebellion against the proprietary patent, Warwick was hoping that they would admit him and the committee of foreign plantations as the overlords of Barbados.

This attack on the proprietorship did not go unnoticed by Carlisle. The earl viewed the ordinance of Parliament not only as an attempt to undermine the allegiance to the king in the colonies but also as a scheme to dispossess him of his West Indian holdings and to "invest my Lord of Warwick into my right and inheritance in the Barbados."[19] So unnerved was Carlisle by the statute that he persuaded the king to issue a proclamation instructing the islanders not to recognize Warwick's commission.[20] Carlisle also sent his personal secretary to the island with a letter to the governor and the Council. In his communication, Carlisle demanded that the colonists not "admit or give way to any (of what pretence so ever, much less to that earl now in rebellion and proclaimed a traitor by his majesty) into that right and inheritance which (you cannot but know) I have in these islands." If the planters refused to heed his words, Carlisle forecasted "bloodshed, rapine, and ruin" among them. He finally assured the settlers that he was resolved to sail to the colony and, upon his arrival, would confirm their "estates and affairs of the island" to the contentment of all.[21]

In a private communication to Bell, Carlisle gave notice that Warwick was sending over "one Raymond." Raymond, according to Carlisle, was coming to the island to "sow faction and sedition" in order to prepare the way "for letting in of one Mr. Humphrey (a New England man) to be governor." Carlisle therefore expected Bell's cooperation "in a business of so great consequence and concern to us all."[22]

In the midst of this growing competition for the loyalty of the colonists, apparently in hopes of defusing the situation, the trustees gave an answer to the Assembly's September propositions. Actually, the trustees responded to only two of the more than thirteen articles presented them. They agreed that the arrears should be used for public purposes, and they also promised to confirm the inhabitants in their lands if the planters would pay a rent of eighteen pence per acre rather than the three pounds of cotton or tobacco proposed by Carlisle. The reason for this change, the trustees explained, was because cotton and tobacco were "very uncertain commodities," and seeing that the colonists had "fallen upon the planting of sugar," it seemed fair that a fixed monetary rate for the rent should be established.[23] But with Warwick suggesting freedom from all taxation, the proprietor's newest offer was destined to get a cool reception in Barbados.

While the landlords continued their struggle to confound the "several designs" upon Barbados in London, the earl of Marlborough began yet another move to wrest control of the "Caribbee Islands." In the summer of 1644, Carlisle defected to Parliament, and in his absence from court Marlborough told Charles that Carlisle had sold the West Indian colonies to Warwick. Since Warwick was "a known and proclaimed rebel and traitor," Marlborough argued, the islands reverted back to the crown by forfeiture. By these means, he acquired a royal patent for the islands. The earl then sent Capt. Thomas Ellis, an old member of the plantocracy, to the colony with a commission to be his deputy governor until his arrival in the Caribbean.[24]

With overtures from Warwick (1643), the proprietors (1644), and Marlborough (1645), Barbadians were faced with a number of critical decisions upon which the continued peace and prosperity of the island depended.

When one looks at the colony's political community in the 1640s, it is immediately clear that there was no broad ideological consensus that might have fastened it to the cause of Parliament or to that of the king. The sugar revolution attracted Catholics, Anglicans, Congregationalists, Presbyterians, Familists (a radical sectarian group), Royalists, and Parliamentarians to the island.[25] Nor was there universal support for one of the would-be landlords. Some householders were for Carlisle; others were for Warwick; there were followers of the trustees; a few were for Marlborough; and the average planter disliked all proprietors. Rather than associate with one of the warring factions in England, the colonists adopted a centrist policy that subordinated all issues to the test of what was good for the security and profitability of the island's plantation system. The existence of this collective agreement is clearly visible in the religious settlement that was formulated.

They granted toleration to all those whose religious activities did not jeopardize the social and political equilibrium of plantation society. As the perceptive Father Biet put it, religious freedom was given to everyone, "provided that they did nothing to be conspicuous in public."[26] The Puritan minister James Parker was of the same impression. Barbadians, Parker wrote, "were unwilling to press any man's conscience in matters of ceremony, only in regard that there are many sects under pretence of liberty may take occasion to deny all ordinance[s]."[27]

The suppression of radical sects was not rooted in a zealous commitment to a religious orthodoxy. The island's political community was composed of Catholics, Congregationalists, Anglicans, and Presbyterians. It would have been difficult for this motley body to agree on a dogma. The hostility toward the radicals was brought on by the potential danger that the spread of sectarianism posed for plantation society, and by the refusal of the radicals to "be contented with enjoying their own ways and opinions in a private manner." Instead, they went about "preaching and maintaining their own conscience and opinion everywhere," thereby breaching the religious compromise the planters had formulated.[28]

Householders in the colony took a similar stance on political division. By an act of the Assembly it was ordered that "every dispute or argument in the colony between adherents of King and Parliament shall be highly punished so that hereby everything is tied."[29] No one was permitted to disturb the public peace because of affections for an English party. Indeed, so eager were sugar producers to avoid the disruptive forces of the English conflict that they fostered a local custom in which "whosoever named the word Roundhead or Cavalier, should give to all those that heard of him, a shoat and turkey to be eaten at his house that made the forfeiture."[30]

In this same light, the colonists behaved in full knowledge that acceptance of a commission from either Warwick or Marlborough would be construed in some quarters as a declaration for an English faction, whatever the personal motivations of the earls were. Such an affiliation with an English party, the planters rightly surmised, would bring "a ruin" upon the island and "destroy it in its beginning to flourish."[31] The settlers also realized that, by staying aloof from the controversy in the metropolis, they could achieve a greater degree of self-determination and freedom from external interference. Although the islanders clamored for Warwick in 1638, in May 1644 they answered the earl and the committee of foreign plantations with polite equivocations, expressing an unwillingness to alter their present government and thus lose the safety of their neutrality.[32] The "fair and large promises" of Warwick for release from payment of all arrears and future rents did create a desire among planters to be free of all rents. The Assembly passed an act sometime in 1644 ending all negotiations with the proprietors.[33] Archibald Hay's nephew William Powry wrote from the island that the landlords need not expect anything further from the colony, unless Carlisle came to the island "in person and strongly too."[34] Thus proprietary government came to an end in Barbados, except for a brief moment between 1650 and 1652. As for Marlborough, Captain Ellis was prevented from publishing his commission, and when the earl arrived in August 1645 the planters refused to admit him. Marlborough thereupon left the island.[35] Through their neutralist policy the colonists escaped involvement in the English problems, and they also managed to avoid direct subordination to any agency in England. The firmness of their resolve was tested again by Warwick.

In September, Carlisle obtained a decree from the House of Lords reinstating him as the proprietor of the Caribbee Islands, "notwithstanding anything that may be objected to the contrary."[36] By this action the Lords created a parliamentary challenge to Warwick's authority over the islands. No doubt conscious of this development, Warwick was able to get the November 1643 ordinance reissued by Parliament on March 21, 1646. It affirmed his right to govern the West Indian colonies.[37] Armed with this rejuvenated power, Warwick made an all-out military assault on Barbados.

From as early as 1638 Warwick kept retainers in Barbados. His chief servant on the island was Capt. James Futter. It was Futter who helped to construct the faction that fervently supported Warwick's proprietorship in the tumultuous years of the late 1630s, and he was still busy in Warwick's service in the 1640s. In fact, sometime in 1644, perhaps around the time the Assembly rejected Warwick's initial advances, Futter was again imprisoned.[38] Warwick came to understand that Futter received such hard measures "because his employment and affairs had some relation" to his interest. Warwick cautioned Bell to make sure that justice was done Futter so that the "Parliament may be saved the trouble."[39]

In preparation for a second attempt to have his commission enforced in Barbados, in the summer of 1646 Warwick sent Sergeant Major Raymond to the colony. This was undoubtedly the same Raymond about whom Carlisle had warned Bell of in 1644. Warwick requested that the governor afford Raymond that "entertainment and countenance which may become a gentleman employed by me and others of quality."[40] He also asked that if Raymond decided to remain in Barbados he be granted some "employment which may suit an approved soldier." In the same communication Warwick announced the coming of another commission from Parliament and expressed a wish that Bell speak in its behalf.[41]

Bell had no intentions of disturbing the equilibrium upon which his authority was based. It did not take long, therefore, before Raymond and the Warwick faction were at loggerheads with the governor. There is no information about the exact nature of the controversy, but we do know that the reports of the conflict supplied Warwick with ammunition to use against the colonists. The earl was told by Raymond and others that they lost their employment in Barbados because they were friends of Parliament.[42] The governor general, in turn, wrote a strident letter to Bell giving vent to his indignation at the governor's refusal to submit to that authority that "the Parliament out of their affection for the common good" of the colony had granted to him. He also predicted that inconveniences would follow if Barbadians "molest or discountenance any person inhabiting there under the notion of Parliament's friends." He therefore required that anyone dismissed from their employment or otherwise troubled in their estate for supporting Parliament be immediately restored to their former place. How well the colonists complied with his demands he would use as a measure of that "love of peace and justice" that the islanders constantly argued they were aiming at in their "waiving of all declarations on either side in the public quarrels" of England.[43]

Before receiving a reply to his letter, Warwick on March 27, 1647, forwarded his strengthened commission to Barbados with a second invitation to the planters to accept his domination. He also empowered some persons on the island to examine "some grievances" in the colony. Apparently, Warwick was hoping to rally another faction that would give him control over the island. Finally, Warwick "authorised . . . masters of ships to assist such in those islands as should declare for Parliament, and to deny trade with such as had fallen to the enemy."[44] Had Warwick made such an assault on Barbados in the 1630s, he surely would have been abetted by a powerful local faction. But, in the environment engendered by the English Civil War, planters were resolutely determined to remain neutral and free.

James Parker, a Puritan minister in the colony, commented that Warwick's warrant to call the governor and Council into "question for some pretended acts of injustice" had gone to "inferior persons here, and not of the best in matters of opinion, and that seems to exasperate."[45] After some agitation, planters were able to subdue Warwick's small party, and in October 1646 the Assembly sent another rejection to the earl. Warwick was respectfully made to understand that, while the island would receive with favor traders from ports controlled by Parliament, it had been decided by "a general declaration of the inhabitants . . . not to receive an alteration of the government, until God shall be so merciful unto us as to unite the king and Parliament."[46]

Carlisle took heart from Warwick's defeat and prepared to go to the colony himself. The prospects of Carlisle in the West Indies was unwelcome news to many. On January 8, 1647, the creditors of the deceased first earl of Carlisle petitioned Warwick and the committee of foreign plantations to stop Carlisle's departure, which they claimed would ensure that they were not repaid as the first earl intended.[47] Some merchants and planters of Barbados also petitioned the committee that their lands be "settled in free and common soccage."[48] On January 11 the House of Lords moved that Carlisle's decision to go to the West Indies be referred to Warwick and the committee of foreign plantations, a rather interesting morsel to set before Warwick.[49]

On January 29 the committee asked Carlisle to respond to the two petitions, and he was also given a copy of the ordinance of 1643 that made Warwick governor in chief of the English Americas. Carlisle answered that he intended to pay his father's creditors, his departure notwithstanding. He further maintained that his reason for going to Barbados was to settle the question of land tenure; hence, the merchants and planters had no legitimate grounds of complaint. As for the ordinance of Parliament, the earl argued, unperturbed by the proper sequence of events, that it was "never intended I should be prejudiced . . . in my inheritance" by the statute but "rather receive assistance and protection by it against the Earl of Marlborough, or any other person, who by commission from the king did or might disturb me in my government and possession."[50] Warwick's authority therefore did not overrule his.

Whatever the merits of Carlisle's case, the decision of the committee was a foregone conclusion. Carlisle, sensing his vulnerability, sought the aid of Francis Lord Willoughby, a prominent member of the Peace party in the House of Lords who was to become temporary speaker of the House in July 1647.[51] In return for his protection, on February 18 Carlisle leased Willoughby the Caribee Islands for twenty-one years. Carlisle also made Willoughby governor general over these West Indian islands for the same period of time.[52]

As expected, on March 1 the committee ordered that it be reported to Parliament that Carlisle's departure to the Indies would "tend to the disturbance thereof, the hindrance of trade, and the discouragement of the planters, and will endanger the defection of the planters from the Parliament."[53] Warwick himself read the committee's report to the peers. In spite of the committee's finding and Warwick's presence, Willoughby was able to carry the day for Carlisle; consequently, the Lords overruled the committee and granted Carlisle permission to travel to the colonies.[54] Two days later, however, the House of Commons denied Carlisle the right to leave the kingdom. The Commons also empowered the committee of foreign plantations to send for Carlisle's patent and the 1643 commission to Warwick, and to consider them both in light of an ordinance prohibiting "delinquents to have any place of trust."[55] This committee, dominated as it was by Warwick, was to report its conclusions to the Commons with recommendations on what should be done with the Carlislian patent.

Just as a parliamentary battle over the Caribbee Islands seemed to be taking shape, the whole matter was dropped. On May 13 the Commons did order that the committee of foreign plantations be revived to "move the business concerning the Earl of Carlisle's Letters Patent," but nothing came of it. In all likelihood, it was the radicalization of English politics in the summer of 1647 that led to the impeachment of Willoughby and the retirement of Warwick that brought the issue to an unresolved conclusion.[56]

During the conflict in England, Barbados stood independent from direct subordination to the mother country. This independence, however, was a precarious one, predicated more on political maneuvering and chaos in the metropolis than on the vitality of the island's social organization. Indeed, inherent divisions in the island's slave plantation system helped to pave the way for the return of imperial rule.

CHAPTER 7

A War for Home Rule

"We are but a month from you by sea," Peter Beckford, a Jamaican planter, mused in a letter to a friend, "but those who talk of a new world of the Indies say more than they oft-times know." Englishmen in Barbados would have agreed with Beckford's observation. It was impossible to confuse a plantation village (fields of sugarcane, big house, and slave huts) with an English manor or farm. In great measure, this new world of the Indies was a product of planter expectations. English pioneers who made the errand into the Caribbean wilderness did not see themselves as part of a Quaker "holy experiment"; nor did they commit themselves to a Puritan vision of a "city on the hill." Theirs was a material dream, a desire for wealth that was not a by-product of Protestant asceticism, which counseled abstinence and subordination of economic activity to Christian values; it was an unabashed materialism in which individual profit and interest were the highest social ideals. In this competitive arena, only a tiny minority were able to fulfill their ambitions. In the main, they succeeded because they came to the colony blessed with social and financial advantages, though these by no means guaranteed success. Among those planters who climbed to the top of the Barbadian social order there was a common and deep pride in accomplishment. In the space of forty years, they had harnessed nature's and man's energies to transform an uninhabited tropical island into a booming colonial center that fed raw materials into a world market system.

Around the island's plantation economy flourished a society that reflected the credit and debit columns in planters' accounting books. It was founded on a lopsided distribution of wealth that transferred much of the community's resources to a small group of householders, leaving the average Barbadian in a state of perpetual poverty. In trying to maintain the stability of plantation society while simultaneously preserving its basic inequalities, planters became enmeshed in a set of social relations, profoundly different from those they were accustomed to in England, that eventually distorted their social outlook and compromised their political values. In the political controversy and civil war that erupted between Barbados and England over home rule in the period 1650–52, planters learned the bitter lesson that the internal dynamics of the newly organized slave plantation system placed limitations on their political behavior. They had to accept the colonial relationship as the price of protection from the social upheaval that awaited an independent Barbados.

The execution of the king by Parliament in January 1649 sent a shock wave through the Barbadian planter class, as it announced the return of a strong central government in England. Not all Englishmen in Barbados were enchanted with the colony's policy of neutrality. A small clique of large plantation owners who were out of power in the colony, but who had close ties to Parliament and to the London merchant community (James Drax, Thomas Noel, and John Bayes are notable examples),[1] could realistically anticipate officeholding in Barbados once Parliament assumed control over the island. Thus, after the defeat of the king, they began to challenge the colony's governing consensus. Although they were in effect pushing for a revitalization of the colonial relationship, it would be wrong to describe these men as being ideologically committed to furthering the ends of English imperialism. They looked to the mother country as a power capable of putting them in office in Barbados. Once in authority, their ambition was to "hold a correspondence (as with other strangers) in commerce and trade" with the metropolis but to minimize Parliament's interference in the island's daily affairs.[2] Even John Bayes, who was most eager to promote closer ties between the colony and the mother country, did not want to see Barbados completely subordinated to England. Repeatedly he recommended that the colony be given two seats in Parliament, where the planters might speak on matters that touched their interest.[3]

More ideologically bound to Parliament were several groups of radical Protestants on the island. A few large plantation owners can be identified as members of radical sects, but sectarianism was most common among poor planters, propertyless freemen, and indentured servants. It was more than coincidence that the sects began to attract an increasing number of the island's poor after 1643. The transition to sugar cultivation and the mass importation of slaves was making some rich, but for most white planters (tenants and small landholders) the developing slave plantation system promised little opportunity for employment and no hopes of material benefit. For some poor whites religion became the language through which they argued for fundamental changes in the emerging structure of Barbadian life. In Parliament's victory, these sectarians saw the beginnings of the social revolution they longed for in the colony, and they were eager to promote a parliamentary takeover of the island.

The majority of Barbadian plantation owners would have been perfectly happy to remain aloof from English party politics and enjoy the benefits of independence. These planters had grown accustomed to ruling the island without deferring to England. They welcomed the fact that legal disputes could be resolved in Barbados without appeal to the mother country, and they appreciated the advantages of having local officials who were directly accountable to them. The island prospered greatly by the free trade it conducted during the turmoil in the mother country, and sugar producers were deathly afraid that through political maneuvering in England some individuals might be able to establish a monopoly over the colony's market. Even more frightening for all householders, a resuscitation of the imperial system under Parliament could have led to the return of proprietary rule, raising unresolved questions about land tenure and rents. With support from England, Carlisle, or any other proprietary power appointed by Parliament, could have summarily dispossessed planters of their landed estates.

The members of the Barbadian Council and Assembly shared the general misgivings about the subordination of the island to Parliament. In addition, they were convinced that the reorganization of the imperial system would result in their removal from office, particularly since many of them had resisted Parliament in the past. The few Royalist exiles (Humphrey and George Walrond, Thomas Modyford, and William Byam) who managed to become members of the government during the period of neutrality were especially outspoken in their criticism of a revitalization of the colonial relationship under Parliament. These émigrés realized that their sway would be negligible with the power that they fought against at home; hence, their dismissal from office and the confiscation of their plantations were a very real possibility.

By the time a unified central government reappeared in England in 1649, many colonists, for various reasons, were openly advocating home rule. Even the most temperate planters were persuaded that the island should be treated as if it were an English town or corporation with the right to elect its own officials and to elect representatives to Parliament. On this latter point, in 1651 the island's General Assembly boldly declared that being "bound to the government and lordship of a Parliament in which we would have no representatives, or persons chosen by us . . . would be a slavery far exceeding all that the English nation hath yet suffered."[4] When put to it, as they soon would be, many Anglo-Barbadians were quite willing to risk open confrontation with Parliament rather than submit to the reestablishment of effective imperial rule. The island's plantation community, however, was not up to the task of preserving the freedom that had been gained in the 1640s. As one planter put it, Barbadians, "having not attained to such a perfecting as to subsist of themselves, must be content to submit to England." But, "if they had the power to dispute it," he concluded, they would have "declared themselves a free state."[5]

The prevailing view in the mother country was that England alone should enjoy the markets of her colonies. This notion was rooted in a mercantilistic view of empire. The idea was that the metropolis would supply the colonies with food, clothing, and equipment while monopolizing those colonial commodities that the mother country could not produce itself. By these means, England would be spared from trading with foreigners for needed raw materials, and thus it would achieve self-sufficiency. To realize this goal, England was quite prepared to direct the internal affairs of its possessions, and it could be anticipated that when central authority was strengthened in the mother country there would be a determined effort to bring the islanders into line with English expectations.

On the day Charles was put to death, Parliament declared it treasonous for the American colonies to proclaim anyone king of England or the dominions without its prior consent.[6] In February 1649 it made the Council of State (which by the same act became the executive branch of government) overseer of the English Americas. The Council of State, however, hampered by other, more pressing matters, did little to impose its authority on the colonies in the first year of the Commonwealth. Indeed, it was not until July that the Council of State got around to sending letters to formally notify the American colonists of the changes in the English government.[7] Aware of Barbados's past obstinacy, on March 15 the Council of State asked a committee of merchants to investigate whether it would be advisable to allow the transportation of horses to the island in case of the colony's disaffection to the Westminister government. The merchants apparently found conditions in Barbados to their satisfaction; at least they recommended that permission be granted to those who wanted to ship horses to the island. Finding no cause for dissent, in May 1649 the Council of State authorized Maurice Thompson and William Penoyre, two prominent London merchants, to export fifty draft horses to Barbados.[8] Although this action by the Council of State is sometimes taken as confirmation that relations between Barbados and England were amicable, it is important to remember that they disagreed over the question of sovereignty in the colony. As it demonstrated in 1643, 1646, and again in 1649, Parliament maintained that authority in the colony emanated from it; Barbadians on several occasions refused to accept this claim. The issue was reopened by the Council of State in its July communication.

While these letters, as George Beer long ago noted, were a "clear enunciation of the formulation which summed up the reciprocal duties of the metropolis and dependency," the colonists being enjoined to "maintain their obedience, as they looked for protection from England," they required no explicit action on the part of the settlers to signify their compliance to the new government, leaving planters free to pursue former policies, though they were put on notice.[9] Through a series of tangled events planters were soon at war with the Commons—waging a hopeless struggle to protect the autonomy they had come to enjoy. It is ironic that their failure to preserve home rule was related to their decision to use slave labor in their households.

The renewed demands of Parliament raised great apprehensions in Barbados. Some of the more determined proponents of home rule, led by Humphrey Walrond, thought that the colony should begin to prepare itself for the moment when it would have to defend its independence. Accordingly, they warmly endorsed a proposal from Bermuda that the two colonies form an "offensive and defensive league" whose underlying purpose was to resist Parliament.[10]

Humphrey Walrond was the most outspoken advocate of the Bermudian alliance. He contended that it would put Barbados in a strong bargaining position with the dethroned heir, Charles II. Indeed, as things stood, he observed, "it was easy to obtain anything from the king."[11] The major difficulty with the union, as more moderate planters correctly perceived, was that it broke with the proven policy of neutrality and committed Barbados to certain hostility with Parliament. Although vehemently advocated by Walrond, these latter considerations prevailed, and the defensive and offensive league with Bermuda was rejected.[12]

Shortly after the departure of the Bermudian envoy, Barbados was embroiled in another crisis. Despite the growing reliance upon servants to help them govern the large number of slaves in their households, planters still kept their Christian bondsmen in a state of deprivation and dependency. Perhaps inspired by news of Parliament's execution of the king, a band of white servants no longer willing to tolerate the conditions under which they lived vowed to "break through it, or die in the attempt." Before they could set their plan into motion, the conspiracy was unearthed. In suppressing the revolt, planters executed eighteen of the principal conspirators, because "they found these so haughty in their resolutions, and so incorrigible, as they were like enough to become actors in a second plot." As an additional precaution, planters decided to keep a "special eye" over the surviving participants.[13]

In this atmosphere, encouraged by Humphrey Walrond, a general paranoia seized the plantocracy. They believed that radical sectarians, abetted by unscrupulous place seekers (office seekers), were conspiring with Parliament to undermine the island's independence.[14] This fear convinced the General Assembly to pass an "Act and an Oath." The act decreed that anyone who by "deeds or words maliciously deprave, vilify, or oppose" the government was to be adjudged an enemy of the island.[15]

Rather than helping to promote political stability on the island, the bills produced charges that the General Assembly, at the instigation of Walrond and his friends, fabricated the conspiracy to conceal a scheme to transform the sitting General Assembly into the permanent rulers of the island. The statutes, many complained, "tended to the perpetuating of them [the Council and the Assembly] and theirs in a place of authority; and to the setting up of a government of will and power."[16] Throughout the island petitions were collected that called for new elections. To the dismay of the members of the Council and the Assembly, Governor Bell agreed to the demands of the petitioners.[17] Unwilling to see elections held in this hostile climate, the Council and the Assembly rose up in rebellion, declaring that the leaders of the petition campaign were actually agents of Parliament who had instructions to foment divisions between the governor and the Council "whereby our enslaving may the more easily be facilitated." Everyone, they argued, should vindicate the Council and Assembly and "labour to prosecute to the terror of those instruments of this evil amongst us."[18]

In less than a week, Walrond had frightened a majority of the planters into believing that friends of Parliament had a plan and that "they were the originals of the petitions."[19]

With a large and credulous faction behind him, Walrond sensed that he might establish himself as the political leader of Barbados. Bell was painfully cognizant of Walrond's ambitions and realized that he might soon be the subject of an attack. On April 29 the governor issued a proclamation that no one should take any arms, or act in any hostile manner, upon pain of death.[20] The edict, however, did not deter Walrond.

On April 30, Walrond persuaded a "plain, but over credulous" Col. Henry Shelly (a Royalist refugee) and Col. Edmund Read (an old planter who had sat on the Council since the 1630s) that agents of Parliament had captured the governor and were plotting to seize one of the island's forts. The colonels raised their respective regiments and advanced against the governor.[21]

Upon receipt of information that he was to be attacked, Bell issued a commission to Lt. Col. James Drax, a leader of the petition campaign, to raise the militia for the preservation of the peace. Drax was able to capture a son of Walrond, "but so inconsiderable were the governor's forces" that Bell was obliged to sue for peace on May 1, one day after Walrond gave the alarm.[22]

The Walrondites placed heavy demands on the governor, including the acceptance of their protection and the bringing to trial of twenty planters whom they chose to nominate. In less than four years after his arrival in Barbados, Walrond was on the verge of assuming political control over the island, but just as he drew near the seat of power fate intervened in the person of Francis Lord Willoughby.[23]

After his flight from London in 1647, Willoughby joined the service of the king; in December 1649 his estate was confiscated by Parliament. "Since all was gone at home," Willoughby later reflected to his wife, "it is time to provide elsewhere for a being," and so he turned his attention to Barbados. During his attendance at Charles's court, Willoughby was able to have Marlborough's commission for the governor generalship of the Caribbee Islands revoked, and his own commission from Carlisle confirmed in its stead. Armed with the blessings of Carlisle and Charles, Willoughby sailed to Barbados, where he arrived quietly on April 29, the day of Bell's proclamation.[24]

Ignorant of Willoughby's arrival, and quick to spot an opportunity where an advantage might be gained, Thomas Modyford advised the entrapped Bell that he could break the grip of the Walronds if he were permitted to raise the Windward Regiment. Floundering hopelessly, Bell agreed to Modyford's plan for his restoration. Modyford thereupon sent word to Lt. Col. John Birch, an old planter who had served in either the Council or the Assembly since 1630, that Walrond had raised an army and surprised Bell. Birch was commanded to call out his regiment (1,500 foot and 120 horse) to force the governor's release. Walrond quickly came to terms with Modyford, who treacherously had Birch's troops disbanded, leaving Bell a prisoner.[25]

Carefully watching all that transpired, Willoughby sent a message to the General Assembly of his new commission. On the surface, to receive Willoughby meant to recognize the king and the proprietor—one a deviation from neutrality, the other a return to proprietary rule. In the eyes of Barbadian planters, however, acceptance of Willoughby had other, more appealing features. The attractiveness of Willoughby increased, for instance, as the General Assembly considered what further disorders might occur if things were left in their present condition. A glance at their frustrated slaves and at the indentured servants and poor whites in their midst unfolded in one nightmarish moment the true horror that civil strife could unleash in plantation society. Since Willoughby intended to reside in the colony, he would be accessible to the planters and would have firsthand knowledge of the society's needs, and therefore would not present the disadvantages of an absentee overlord. In this regard, Willoughby was to be more than a proprietary governor of Barbados. He leased the islands from the proprietor and thus could rule without Carlisle's interference. In fact, during the course of the negotiations with the General Assembly, Willoughby agreed to confirm householders in their estates in return for a duty on all exports from the colony. Furthermore, Willoughby promised to obtain a commission from Parliament. If this latter was accomplished, the colony's independence might be secured for the future. On May 7, 1650, a majority of the General Assembly voted to receive Willoughby as the new governor, not out of an addiction to royalism or a desire to reinstitute proprietary rule, but out of the hope that Willoughby could return the island to its former peace and preserve home rule.[26]

Some colonists were opposed to the new governor. A handful of planters, apparently members of the defeated petitioner coalition, tried without success to interrupt the negotiations between Willoughby and the General Assembly.[27] Inside the government, Walrond was annoyed at the twist of fate that thwarted him as he was about to claim political leadership of the island, but he and his supporters could not oppose Willoughby without going against their "declared principle of loyalty to their kingly stalking horse." They tried, however, to arouse distrust for Willoughby by saying "that he was once a Roundhead and might be again." By insinuations and influence, Walrond forced a suspension of Willoughby's authority in the colony for three months, during which time the rebellious General Assembly was to remain seated. This was done, it was pleaded, "in respect of the incertainty and distractions of the present times."[28] Walrond presumably argued that this period was needed to eradicate parliamentary agents who, if left unchecked, would soon hand the island over to the Commons.

Even though the General Assembly was sympathetic to the moderation that they saw in Willoughby, they were determined, indeed bound by their accusations, to take action against the petitioners. On May 23 the General Assembly banished and fined 122 persons whom it depicted as "independents, non-conformist to the doctrine and discipline in the Church of England, and others their aiders and abettors in the disturbance of the peace of the island." These were by no means all those who signed the petitions. They were outspoken sectarians who challenged the legitimacy of plantation society on religious grounds. Also deported were those whose close connections with the London merchant community and Parliament made them appear as place seekers. Strikingly, in spite of the fact that several large planters, including James Drax, Thomas Middleton, James Futter, and Thomas Noel, were ousted from the colony, few of the exiled petitioners had ever sat in the General Assembly, none more than twice. The deported colonists were not the traditional political elite, as they are often characterized, and it was not their removal from power that started the civil war. They were the men and women whose continued presence in Barbados, in the opinion of the General Assembly, threatened the survival of the island's independence.[29]

Whatever the beliefs of the Barbadian Council and Assembly, in the main, deported planters were not eager to promote a colonial system in which Parliament would minutely direct the affairs of the colony. It is possible that some sectarians entertained such dreams out of hopes that it might translate into a restructuring of the society, but as far as large estate owners who were deported are concerned, on the issue of the long-term organization of Anglo-Barbadian relations, they were in common agreement with other members of the plantocracy. All wanted Barbados to be incorporated as an English town, with the right to elect local officials and to elect representatives to Parliament.[30] Even John Bayes, who later was to argue for a more centralized method of governing the colony, at the time of his banishment argued for this type of union.[31] What exiled big planters wanted, which put them into conflict with the Council and the Assembly, was for Parliament to put them in power in the colony. Beyond that, they were resolved to minimize imperial interference in the island they hoped to govern.

Their immediate concern was to move the English government to act on their behalf in Barbados, and once in England they actively campaigned for assistance.[32] On August 27 they petitioned Parliament to declare all householders remaining in the colony rebels and traitors. They also requested that a fleet be sent against them and advised that a new governor be appointed immediately.[33] The purpose of all these demands, one hostile observer commented, was to force the island "to accept of such conditions as they which command the fleet think fit."[34]

Several London merchants, supporters of Willoughby, were strongly opposed to the petitioners' belligerent supplication. To proclaim the Barbadians rebels, the merchants remonstrated, would drive the islanders to extreme measures, rather than allow themselves to "fall into the hands of those whom they had provoked, who threatened them with fire and sword at their departure, who must now become their judges and afterwards as far as they please their masters."[35]

The merchants' cautions were well founded. While maneuvering to secure his position in Barbados, Willoughby was apparently able to get word to several merchants, informing them of the situation in Barbados and of what would happen if the colony were attacked by England. It seems that his message to these merchants was incorporated into an anonymous pamphlet written to counter the proposals of the refugees.[36] Willoughby's behavior, the pamphlet began, was to be "understood as one that hath the helm given into his hands in a storm that he is as yet fair to steer according to the temper of the wind and seas." He was therefore to be excused for being unable to stop the banishing of those who were "cast out the ship for its preservation." The colonists, the essayist surmised, had not "attained to such a perfecting as to subsist of themselves"; consequently, they "must be content to submit to England." However, if they had the power to win a dispute with the mother country, the pamphlet menacingly went on, they would have "declared themselves a free state, instead of proclaiming a king."[37] The people of Barbados, the author argued in anticipation of a declaration the islanders were to make a few months later, left England as "a free people to dig out their own fortunes in a strange land." The settlers had improved the island at their own charge without public expense, and "never had too much protection" from England. "But it is without question that the proper interest" of Barbados, the pamphlet concluded, "is still to stick to England under what government soever it be." If the island were attacked, however—the tract returned to a more belligerent tone—the estates of merchants would be seized, and indentured servants would be liberated and put "into possession of the sequestered [estates] which with easy husbandry may be brought to offer them subsistence though not wealth." The essay even dared to suggest that the slave population, valued at over £500,000, could be sold to the Spanish. In short, the pamphlet suggested that social reforms would be undertaken in Barbados that would give indentured servants and poor whites some incentive to fight for the social order and eliminate the possibility of a slave insurrection. Through this argument Willoughby reached a portion of the London merchant community and persuaded them that an assault on Barbados would be costly.[38]

Parliament was not impressed with these threats and rejected the merchants' petition. On October 3 it passed an act halting all commerce with Barbados and empowered the committee of the navy to prepare a fleet to subdue the colony.[39]

With so much at stake, the merchants were unwilling to concede the issue as lost. Upon completing negotiations with Lord Willoughby's agent (George Marten, brother of the regicidist Henry) in England, on November 22 the merchants petitioned the Council of State for permission to trade with Barbados if the colony would subscribe to four propositions that would permit the exiled planters to return to the island in peace but that would allow Willoughby and the General Assembly to continue in power.[40]

The most vocal critic of the merchant compromise was John Bayes. The present government in Barbados, he objected, "were the chief and principal actors in the late rebellion," and he would not feel safe under the rule of such people. Besides, Bayes had learned from conversations with Marten that the "powers that do now govern in Barbados were not sensible that they had done anything against this commonwealth"; rather, they "looked upon themselves as a free people."[41]

In an unsigned petition, which bore the influence of Bayes, several Barbadians demanded that Willoughby be replaced. They charged that the governor had abetted the rebels; thus, he should not be continued in office. The petitioners then went on to sketch out their thoughts on the future of Anglo-Barbadian relations. "As freeborn Englishmen," they wrote, "we desire in Barbados that as all power in all places here in England do receive their immediate commissions for the exercise of all authority from the High Court of Parliament which is the representative of the whole nation, so Barbados as a branch belonging to this commonwealth may be entirely incorporated into the same as any town, city, shire, or island thereto belongeth."[42] The effect of their proposal would have been to secure virtual home rule in Barbados.

The Council of State did not respond to the latter request of the colonists, but it did dismiss the merchants' compromise as being "dishonorable to [the] Council" and thus vetoed any possibility of Willoughby remaining as governor. Sir George Ayscue, Daniel Searle, and Michael Pack were appointed commissioners by the Council of State charged with the responsibility of going out to the Caribbean and to bring about the reduction of Barbados. Ayscue was made governor with power to choose six councilors, one of whom was to be Searle, who was to succeed him as governor when Ayscue left the island. After some delay, on August 5, a fleet (consisting of 7 ships and 860 men) was dispatched to the colony.[43] The mission of this tiny armada was to reduce an island inhabited by 38,000 people, who were defended by 7,000 militiamen. Numerical superiority was on the colonial side, but plantation society produced few who would be willing to fight to perpetuate the social order.[44]

The tactics used by the English to defeat the island were outlined by William Hilliard, one of the councilors who had formulated Barbados's neutrality in the early 1640s. He left the island for England sometime around 1647 or 1648. Realizing that major damage would be inflicted on the colony's plantation system if it were openly attacked, Hilliard ruled out a full-scale invasion, recommending that "all [the] tenderness that [is] possible may be used towards the inhabitants."[45] If the colony failed to submit at the first approach of the fleet, the best strategy was to establish a blockade, cutting off all trade and commerce. Raids were to be conducted day and night to harass the population. In the end, Hilliard predicted, some planters would adhere to Parliament out of fear of what might happen if they did not, and the rest would surrender if given an act of oblivion.

While events took their course in England, Willoughby feverishly labored to consolidate his position. The restrictions placed on his powers were lifted in August 1650, and he finally assumed the governorship of Barbados with full authority. Upon taking the helm, Willoughby promptly prorogued the General Assembly and held new elections. The men who returned were the traditional political elites; most of them had served in the previous Council and Assembly, except for Humphrey and George Walrond, who were not reappointed.[46]

Matters stood in abeyance in Barbados until early February 1651, when reports from the mother country made it evident that Parliament was determined to enforce the embargo imposed under the act of October 3, and was preparing a fleet to demand obedience from the colony. Wild rumors reached Willoughby that his agent, George Marten, was nearly hanged for speaking publicly in his behalf.[47] Friends of the regime in England wrote the inhabitants to "expect nothing but fire and sword" from Parliament, and "there were to be no terms offered."[48] With a confrontation with Parliament in sight, Willoughby abandoned all vows of moderation. At his insistence, an ordinance was passed by the General Assembly requiring planters to swear an oath to defend Willoughby as governor of the colony against any foe.[49] He had the estates of many banished planters confiscated.[50] Willoughby also spread misleading reports about the intentions of Parliament. In a declaration published on June 11, the governor informed the population that, in spite of the "great moderation that hath been used to convince the exiled petitioners of their errors, and reduce them to their due obedience to this government," he found them "settled against the peace and good of this island." It was also discovered, Willoughby told the planters, that the Council of State had voted to force a governor upon the colony "and a garrison of 1,200 men in arms to be maintained by this country for the government to execute all its commands (how arbitrary or unjust soever)." Willoughby expected that these revelations would provide all men (monarchists and home rulers) with "facts and hands to defend themselves against the slavery that is intended to be imposed on them."[51]

Convinced by this information that hostility with Parliament was imminent, the General Assembly came as close to a declaration of independence as any English colony was to come before 1776. Answering the act prohibiting trade with the island, on February 18 the colonists arrogantly proclaimed that they were the people of England "who with great danger to our persons, and with great charge and trouble, have settled this island in its condition." "Shall we," they belligerently inquired, "be subject to the will and command of those that stayed at home?" "Shall we," the General Assembly asked, echoing the sentiments of many of the deported planters in England, "be bound to the government and lordship of a Parliament in which we have no representatives, or persons chosen by us, for there to propound and consent to what might be needful to us, for as also to oppose and dispute all what would tend to our disadvantage and harm?" "We doubt not," they boldly went on, "but the courage which hath brought us thus far out of our own country, to seek our being and livelihood in this wild country, will maintain us in our freedom." Pushing from their minds the reality of their situation, they estimated that their numbers were not "so contemptible, nor resolution so weak, to be forced or persuaded to so ignoble a submission."[52] With their position enunciated, Barbadian householders collected themselves to defy the power of the mother country.

The task of defending the island fell to the Barbadian militia, but it was a weak and divided institution incapable of mounting a determined resistance to the reestablishment of imperial rule. The militia was a creature of the plantocracy, but it faced real problems recruiting a loyal rank and file from among the island's inhabitants. By 1650 the majority of Barbadians were slaves. Because they knew that deprivation made their slaves unreliable members of the community, Anglo-Barbadians were generally unwilling to induct blacks into the militia where their weight of number might have had unwanted consequences; indeed, one of the duties of the militia was to keep them in a state of awe. "The mustering of men and the reports of their weapons," Richard Ligon observed, "were intended to subjugate blacks to so low a condition as they dare not look up on any bold attempt." For further security, Negroes were not "suffered to touch or handle any weapons."[53] These precautions reduced the possibility of an uprising, but they never completely stopped acts of rebellion. Beauchamp Plantagenet, touring the colonies in 1648, noted that Barbados had "many hundred rebel Negro slaves in the woods."[54] The appearance of a hostile fleet off the Barbadian coast multiplied the hazards for the plantocracy. They could not call upon the bulk of society to man the defenses; and they knew, one islander wrote, that even the most trusted slave at such a time could be "easily wrought upon to betray a country, especially upon the promise of freedom."[55]

Ideally, planters needed a loyal, free population to fill the ranks of the militia. Willoughby probably counted slightly over a third of the islanders in the trainband as fitting that description. For the most part, they were the colony's propertied class (i.e., those with at least one acre of land or one bondsman), though he had to wonder about the extent to which small planters would sacrifice to defend the society. "What heart these poor can have (in case of invasion) to expose their bodies in the way of defence," asked Nicholas Blake, a middling planter who was unusually sympathetic to the plight of Barbadian poor whites, "when victory shall still continue them bondsmen [to their creditors] and overthrow shall set them free?"[56]

Willoughby also had to be uneasy about the presence of propertyless freemen and tenants in the militia. They faced real difficulties in the island's plantation economy. Indeed, many of them chose to migrate from the island.

Large estate owners were fully aware that economic conditions were causing free whites to leave the island to the detriment of the militia. When Willoughby took over the government, he quickly found that many settlers were departing "for want of employment."[57] The anonymous author of "Some Observations on the Island of Barbados" estimated that between 1643 and 1666 more than 14,000 whites left the colony. In his opinion, those that remained were, "for their number and quality, fit only upon a just consequence, rather to betray than defend so valuable a country."[58] While we might quibble with his figures, which are high, the pattern of migration and the attending military repercussions that he sought to document were undoubtedly true.

To fill out the ranks of the militia, planters also inducted indentured servants into the trainband, though not without serious reservations. As has been mentioned, during the height of the sugar revolution, many indentured servants were lured to Barbados by the prospects that they might soon become a master of a great plantation. Once they reached the colony, however, ill treatment and the unpleasant fact that there was little hope of their acquiring any land on the island, let alone a large plantation, made them unruly and restive.

Joining the ranks of discontented servants were a considerable number of Royalist political prisoners. During the course of the political turmoil in the mother country, Parliament sought to rid itself of some of its enemies by deporting them to Barbados, where they were sold as servants. These deportees helped to create a fairly steady supply of indentured servants, but they tended to be highly rebellious. Irish political prisoners are a notable example. Their quarrels with the English reached a new pitch in Barbados. As early as 1644 the General Assembly had passed, to no avail, legislation prohibiting the importation of Irish servants.[59] An intimidated Edward Hollinshead told the Barbadian Council that two of his Irish servants had behaved so "rebelliously and mutinously" toward him and his wife that they were "in fear of their lives."[60] Cornelius Bryan, an Irishman with a menacing appetite, declared while eating meat that "if there was so much English blood in the tray as there was meat, he would eat it."[61] On occasion the Irish joined with blacks to resist their common oppressors.[62] It was obvious that servants given to such bold deeds and sanguinary remarks, as their revolt in 1649 established beyond a doubt, could not be enlisted into the colony's militia without misgivings.

The Barbadian militia, whose ranks were filled with political prisoners, poor whites, and indentured servants, was able to hold off the fleet for three months only because the parliamentary expedition was under strict orders to try to obtain the island's surrender by persuasion rather than by force. The hope in England, particularly among English merchants who traded with the island, was that the colony's sugar plantations would not have to be destroyed in the attempt to demonstrate to planters their dependency upon the imperial connection. Had the fleet been less circumspect about the use of violence, as a truly belligerent foreign foe would have been, the Barbadian plantation system could have been ruined. Appeals to the slave population could have turned half of the society against the social order.

Ayscue and the fleet reached Barbados on October 16, 1651. On

October 17 they summoned the colonists in the name of Parliament. The governor answered that rather than submit to an illegal summons he was prepared to defend the colony against any attack.[63] Following Hilliard's outline, Ayscue began a blockade of the colony and sent ashore small skirmishing parties from the fleet to harass the population. Ayscue also issued a proclamation telling the colonists that the governor had refused to submit to Parliament. Appealing over Willoughby, Ayscue called the planters' attention to the danger of resistance. By opposing Parliament, he told them, the island would be transformed from its flourishing condition into a "seat of war." "Whatever the specious pretences of any amongst you are," he mockingly wrote, "you cannot be ignorant, but that they are altogether unable to give you protection without which this island can no way subsist."[64] In a word, plantation society was dependent on the colonial relationship.

Responding to Ayscue's declaration, the members of the General Assembly announced their intention to risk their lives and fortunes to "defend His Majesty's interest and lawful power in and to this island as also the person of the Right Honourable Francis Lord Willoughby."[65] Strikingly, they made no attempt to refute Ayscue's suggestion that they could not protect the colony.

While the daily raids on the island continued, Ayscue did not attempt a major operation against the island until the arrival of the Parliamentarian fleet bound for Virginia. Momentarily strengthened by the addition of this fleet, and eager to press the psychological advantage that the increased support gave him, Ayscue sent another summons to Willoughby, which the governor again turned down.[66] On December 7, Ayscue dispatched a party of 400 men to attack one of the island's fortifications. This small force routed an entrenched army of 1,200 foot and horse, slaying 100 and capturing 40. In this battle Ayscue suffered only 7 dead.[67] The overwhelming success of this mission, the Parliamentarian party being outnumbered 3 to 1, is indicative of the general morale of the Barbadian militia. In fact, many servants who were needed to fill the ranks of the militia were fleeing to the fleet, where they were promised freedom.[68] The Baconian principle was having its day.

The success of Ayscue's assault provoked serious rumbling among the planters, which caused Willoughby to open talks with the fleet on December 26. Of the eleven articles that Willoughby sent Ayscue, the one to emerge as the critical point of controversy was the first—"that the legal and rightful government of this [island] remain as it is now by law and our own consent established."[69] Upon receipt of Willoughby's propositions, Ayscue drafted fourteen articles of his own. Ayscue was very generous to the islands. He promised free trade with foreigners, no taxation without the consent of the people of Barbados, no garrisoning of English troops in the colony, and an act of oblivion.[70] Ayscue was carefully following Hilliard's advice to pacify the island by presenting the colonists with easy terms. The crucial difference in the fleet's proposal was the first article, which removed Willoughby and the General Assembly from office.[71] On December 29 the Council and the Assembly rejected Ayscue's propositions, answering that "we do unanimously adhere to the first article in our proposition . . . and until a grant first to that, we shall not yield to allow further treaty."[72] The struggle for home rule had been reduced to a battle over place.

Despite the display of unity, the General Assembly was soon deeply divided. At the departure of the fleet from England, the president of the Council of State, John Bradshaw, sent a message to Thomas Modyford through John Bayes. After the rejection of his articles, Ayscue ordered Bayes to deliver Bradshaw's letter to Modyford.[73] Modyford was pleased to receive these "unexpected civilities" that, he later wrote, "sweetly captivated my pride and clearly fixed it in a true affection" to Parliament's interest. Apparently, the president had offered Modyford a council seat in the government that Parliament was about to form in the colony.[74] Rallying to his side others in the General Assembly (notably John Birch, his former confederate in the attack on Walrond; Henry Hawley, the former governor; and Robert Hooper, a large plantation owner), Modyford organized a peace faction that pressed for acceptance of the generous terms that Ayscue was willing to give, which of course meant the abandonment of Willoughby and anyone else who was unable to gain the favor of the new masters of Barbados. Without Modyford's assurances of a future place in the island's government, the majority of the General Assembly refused to surrender to Parliament and charged "that all the desirers of treaty were traitors."[75]

His failure to arrange a submission to Parliament in the General Assembly cast Modyford on a more hazardous course. Encouraged by the securities he had been given, Modyford defected to the Commons on January 5. He also persuaded his regiment to join in this surrender to the Westminster government. Thus the ironies of ironies: it was Modyford, a Royalist refugee, who led the faction that eased the way for the return of imperial rule.

Modyford's plantation became a beachhead for the fleet. Ayscue sent troops to the island that combined with the forces gathering around Modyford to give Parliament an army of 2,000 men.[76] Heavy rains delayed for a few days any contact between the forces of the metropolis and those of the colony. By the time the inclement weather subsided, the extent of the calamity that was about to befall the island convinced Willoughby and the General Assembly that it was time to sue for peace. On January 11, 1652, Barbados surrendered to Ayscue.[77]

CHAPTER 8

The Price of Protection

In the early months of 1652 an occasional planter voice could be heard in support of the articles of surrender, suggesting that the island had secured from the mother country the rights of free trade and home rule.[1] Most plantation owners were far less sanguine about the future of Anglo-Barbadian relations. Although their worst fears were never realized, it soon became apparent that the articles of surrender could not be used to preserve the island's former independence. The capitulation of the plantocracy was followed by a period in which the island was pulled more firmly back into the orbit of England. Planters did not passively acquiesce in English authority; but again and again in specific areas of disagreement about the direction of Anglo-Barbadian relations, in spite of who was in power in England, planters' views always took second place to those of the mother country. Had they had the ability to demand otherwise, it is quite possible that a crisis reaching the proportions of that of 1650–52 would have occurred again. But with a dangerous number of slaves and servants in their households, and a small free population to stand in defense of the social order, to survive in the competitive and contentious environment of the seventeenth century Caribbean planters were obliged to look to England for military protection. England provided military assistance, but at a price. She expected planters to defer to her will. Out of this reciprocal arrangement (protection/obedience) flowed the colonial network. In a real sense, the stability of the imperial system must be understood in relation to the internal dynamics of the plantation household.

In the years immediately after the surrender of the island to Parliament, the planters' discontent with their subordination to England was focused primarily on the office of the governor who, being chosen in England, came to represent the authority of the mother country in the colony. Although the colonists agreed to it in the articles of surrender, shortly after the conquering fleet departed, many began to argue that the governor, being an appointee of the powers that be in England, would be unduly concerned with pressures from the metropolis to the detriment of the colony.[2] Even among those deported planters who returned with the fleet, there was a common unhappiness about having the governor selected by the ruling powers in England. The consensus among plantation owners was that the island should be incorporated with the right to elect its own governor.

In the spring of 1652, inasmuch as Parliament had not yet ratified the articles of surrender, the islanders held out faint hopes that it might be able to persuade the mother country to give in to their demands. Thus, some continued to agitate for the incorporation of the island. Thomas Modyford wrote home that "to demand to have burgesses with you to sit and vote in matters concerning England may seem immodest, but to desire two representatives to be chosen by that island [Barbados] to advise and consent to matters that concern this place, I presume may be both just and necessary." "For if laws are imposed upon us without our personal or implied consent," this master of a large plantation household surmised, "we cannot be accounted better than slaves, which all Englishmen abhor to be, so I am confident you detest to have them."[3]

In England, when the articles of surrender were being reviewed, John Bayes offered written testimony that helped to persuade the Council of State that the power to appoint the governor ought to reside in England. He told the Council of State that a governor chosen by planters would look upon them "as his superiors," and thus "expect their approbation of all his actions." Besides, the "giddy multitude may pitch upon a person unworthy of the trust, who upon flying reports (at this distance) may quit the Parliament's interest." Furthermore, he predicted, in time "should all powers and authority arise from amongst ourselves we shall be as it were another nation and forget our relations to England." In spite of his imperialistic sympathies, he was not totally insensitive to colonial concerns. Bayes reiterated the request that two planters be allowed to "sit and vote in the Parliament of England as representatives of this island, and then there can be no complaints, but that we have a Parliament and as much privilege there as any county in England."[4] The Council of State again ignored the question of the incorporation of Barbados, but it did support the ratification of the articles of surrender, and subsumed in that the selection of the governor by Parliament.[5] While planters failed to make the governor a local appointee, they immediately mounted a campaign to limit the powers of the chief magistrate within Barbadian society, and thereby curb English interference in the daily operation of the government.

During the independent years of the 1640s, the Barbadian Assembly assumed an active role in the governing of the colony. Recognized by the articles of surrender as a branch of the island's government, the Assembly expected to continue its participation in the management of the colony. In the spring of 1652, John Bayes, now a partisan backer of the governor, wrote home that once chosen the members of the lower house "do sit as often as the governor and the Council, and think themselves injured if they do not." Moreover, Bayes complained, the Assembly demanded "an equal (if not the greatest share) in the government, making laws and appointing officers." The Assembly's members were so bold that they had "messages passed between the governor and Council, and themselves like the House of Lords and Commons."[6]

Had the lower house merely desired to act as a third branch of government, the difficulties between it and the governor might not have boiled over into a heated dispute. The Assembly, however, was determined to curtail the powers of the governor, and thus undermine English authority in Barbados. "This parliamentary assembly," as Bayes called it, charged that the present form of government with the governor appointed in England was "too king like" and would soon grow arbitrary. They therefore would not suffer the "governor to judge of any small matters of issue out of his warrant upon any urgent occasion." Furthermore, they demanded that any disagreement that arose over the articles of surrender were to be submitted, not to the governor and Council, but to the island's courts.[7]

In its quarrels with the governor the Assembly found considerable support from among the colonists. The prevailing mood was still for home rule. Reflecting on the articles of surrender, many wondered why they had to accept a governor who was appointed by a distant power. They grumbled that "every corporation in England hath liberty to elect an annual magistrate." Why should they not have the same privilege? Echoing words heard in the lower house, it was protested that "to appoint a governor over them was not freedom but king like." The feeling was that the island should elect its own governor who would be responsive to the colony's needs.[8]

There was also much discontent in Barbados over the fact that the islanders were not permitted freedom of trade, in clear violation of the articles of surrender, which the colonists believed exempted them from the Embargo Act of 1650 and the Navigation Act of 1651. Ostensibly, the Embargo Act was passed to punish planters for their rebellious behavior. A clause in the bill summarized English views on colonial free trade. It plainly stated that in the future all trade by foreign ships to any English colony was to be prohibited, except by special license from the Council of State.[9] Pressured by London merchants, in 1651 Parliament further limited the market where colonials could sell their produce. This ordinance was the first in a series of commercial statutes that were issued in the second half of the seventeenth century that spelled out in practical terms the desire in the mother country to monopolize the trade of its colonies. The act stipulated that no colonial produce was to be imported into England, Ireland, or other colonies except in ships owned, and for the most part manned, by Englishmen, including colonials. In addition, no European goods could be imported into the colonies except in English ships, or in ships belonging to the country where the goods were produced, or to the port of usual first shipment. Colonial trade was not entirely confined to England, for by the statute it was still possible for the colonies to send their produce to the Continent or elsewhere and get manufactures directly from the countries of origin, but later legislation was to close these loopholes.[10]

The governor of Barbados in this period was Daniel Searle. He was jealous of his powers and was quite willing to enlarge his authority at the islanders' expense. For his part, Searle exploited the colonial attempts to restrict his powers in the colony to argue for a broadening of his jurisdiction and the abolishment of the Assembly. In this struggle for power within the Barbadian political establishment the planters' vulnerability quickly became evident.

Searle's difficulties with this restive population would have been simplified if he had had better control over the Council. The governor did not appoint the Council; it was nominated by the English Council of State. Searle had no veto over this body. Some Barbadian councilmen contended that the governor was no more than a chairman, with no more rights than other members of the board. Even more damaging, a few in the Council encouraged the home rule sentiment in the Assembly and in the population at large.[11] Searle therefore could not expect solid support from the Council in any confrontation with planters.

Represented by his wife, in the winter of 1652, Searle approached the Council of State about allowing him to appoint his own council.[12] The Council of State seemed well disposed toward her supplication. On January 5 it ordered the committee of foreign affairs to draft instructions for Searle that would continue him in office for three more years.[13]

In the early months of 1653, Bayes reached England prepared to plead the governor's case. In February he told the committee of foreign affairs that there was a "high faction" in Barbados that was intent upon "lessening the power and authority of the governor" and upon molding the island into a "free state (as they called it) to choose their own governor, establish their own laws, and to have free trade." Bayes warned the committee that if these discussions about home rule were not suppressed, in a short time the people would be alienated from that "due obedience which they are bound to yield to the supreme authority of this government." Bayes recommended an enlargement of the governor's authority; specifically, he advised that Searle's tenure in office be lengthened, that he be allowed to appoint his own council, and that he have an independent source of revenue.[14]

Before the committee could respond to Bayes's petition the Rump Parliament was prorogued by a disgruntled English army. The formation of a new government under Oliver Cromwell, Lord Protector, raised questions about the legitimacy of Searle's rule in Barbados. Some discontented planters argued that the collapse of the Rump Parliament meant that all English authority in the colony had ceased.[15]

It was August 28, 1653, before a commission arrived from the metropolis reappointing Searle governor of the colony. Of great significance, the new warrant empowered him to select his own council.[16] Strengthened by the commission, Searle dismissed three councilmen (Modyford, Birch, and Drax) who were closely associated with the home rule sentiment in the lower house. The governor told the Council of State that their removal was done in order "to put the trust in the hands of such as all along have appeared faithful."[17]

Searle then called for new elections for the Assembly. Overestimating the strength of his new commission, he attempted to influence the selection of the Assembly by putting forth a list of candidates he wanted elected. The results were disastrous. The governor reported that all those nominated by him were defeated, and the people returned men who were "enemies of the commonwealth," many of whom had sat in the rebellious General Assembly of 1650. The displaced Modyford was voted speaker of the lower house.[18]

Even before the sitting of the Assembly, rumors began to circulate about what it might do. Some thought that it would "demand the power of the militia."[19] Fearing their old adversaries, a group of radical sectarians, with the connivance of the governor, sent a remonstrance to Cromwell informing him of the results of the election. Ignoring the act of oblivion included in the articles of surrender, they requested that an act of Parliament against delinquents holding places of public trust be extended to Barbados.[20] Since anyone who was a member of the Council or the Assembly during the crisis years of 1650–52 had been declared a delinquent, the English ordinance, if enforced in Barbados, would have excluded from power many influential planters in the colony. Understandably, the remonstrance drew a hostile response from the Assembly.

Once seated, its members told the governor and the Council that subscribers of that document were trying to make the inhabitants of Barbados odious to Cromwell so as to take away "their undoubted right of freely choosing their representatives."[21] The lower house sent a counterpetition to Cromwell begging that no attention be paid to the misinformation "which some bold insinuating persons at this distance may attempt to give." They also requested that persons sent to govern them be enjoined to observe the articles of surrender. They further asked that "in regard we are Englishmen of as clear and pure extract as any, we may enjoy our part of liberty and freedom equal with the rest of our countrymen and be made proportionable shares of all those blessings which our good God by you his instrument hath bestowed on our nation."[22] While no specific demand was made, it would seem that the Assembly was requesting representation in Parliament.

The Assembly also put before Searle a bill requiring an annual election for the lower house, and the old Assembly "was not to be dissolved until the sitting of the new representatives."[23] The intention was that there would never be an occasion in which the Assembly could not act.

Searle's reply to the Assembly's bill was intentionally provocative, hoping, it seems, to fuel the dispute between him and the lower house in order to argue for an enlargement of his powers in England. Seizing upon the the ambiguous phraseology of their act, he charged that "no representatives of the nation of England in what part so ever of the commonwealth they are in were intrusted with the lives, liberties, and estates of the people, but the supreme authority." To ratify their statute, he maintained, would be to betray the power that was given to him.[24] In a letter to the Council of State, Searle claimed that to accept the act would in effect establish Barbados "as a free state independent to the commonwealth, only to remain under English protection, but not to own England."[25] Before the dispute between the governor and the Assembly could reach a head, there was an abrupt settlement of their differences that was triggered by unsettling developments in England.

Through petitions, official letters, and private communications, the mother country was kept informed of the intense struggle between the governor and the Assembly. This news prompted several London merchants to suggest the restructuring of the Barbadian government in order to more efficiently rule the colony. In January these merchants petitioned the Lord Protector that the distracted condition of Barbados was a threat to their investments on the island. They estimated the Barbadian population at 5,000 freeholders, 5,000 freemen (2,000 English, 2,000 Scots, 1,000 Irish), 8,000 servants, and 20,000 slaves. "By this list," they wrote, "a rational man will quickly judge how difficult it is to keep these several interests peaceably composed under one government." It was also easy to see "what great impression the more necessitous and discontented [meaning servants and slaves] may have on their superiors." With great ease "a desperate, melancholy, and needy soul could improve the discontent of the slaves and servants (which are near alike) so as to wrest this jewel from the dominion of England." This disaffected spirit could marshal the island's servile population to either "set up himself or basely betray it [the colony] to a foreign nation." Even if the rebel failed in his design, if the struggle reached "the heights of blows" the island would be undone, because "their whole wealth consists in servants and slaves who more probably may fall in the encounter."[26] The merchants advised that seven persons should be nominated from England to serve as a council in Barbados. This body would choose one of its members to be governor. A special council would be selected in England to oversee the actions of the Barbadian council.[27] Although the Assembly was not specifically mentioned, the implication was that it would be abolished, and the governor was also to be replaced.

When reports of the merchants' proposals reached Barbados, they brought to an immediate halt the raging controversy between the lower house and the governor. Faced with the permanent abolishment of the General Assembly and with the creation of a West Indian council to minutely govern the island, the Assembly realized that instead of bringing greater independence its defiant behavior was laying the foundations for a more authoritarian form of government that would be sympathetic to, if not under the control of, the English merchant community.[28] Forced by the circumstances of living in a slave plantation society, planters lowered their voices and, for the moment, conceded that the island's governor would be an imperial appointee.

The planters had barely recovered from the shock of the near reorganization of the island's government when they were confronted with a new set of problems that underscored their subordination to England. In the fall of 1654, Cromwell decided to send a fleet to the Americas (under the command of Adm. William Penn and General Venables) in hopes of taking over the vast territories held by Spain in the Caribbean and to establish there a British colonial empire.[29] The Lord Protector was led to believe by Thomas Modyford, who was anxious to ingratiate himself with the new power in the metropolis, that Barbados could help to outfit the fleet with men and supplies.[30]

The expedition arrived at Barbados on January 30, 1655, surprising thirteen Dutch merchantmen. It seized them as violators of the Navigation Act of 1651.[31] Startled by the arrival of the force, the capture of the Dutch ships gave Barbadians a clear warning that imperial rule was going to be sharply more unpleasant while the fleet stood anchored around the island.

Cromwell's "Western Design," as his plan to drive the Spanish out of the Caribbean came to be known, was not a scheme that endeared itself to Barbadians. It meant the establishment of new English colonies in the West Indies. These settlements could become dangerous rivals in sugar manufacturing. They would need people, and there were many unhappy small and middling planters in Barbados who could easily have been enticed to a new settlement. Their departure would have increased the proportion of blacks over whites in the colony, with potentially serious consequences.

Shortly after their arrival, Penn and Venables met with Searle and the Council. The officers were welcomed with civility and were authorized to billet their soldiers with the island's householders, "with an engagement to pay for what they took." After conferring with the general and the admiral, Searle convened the General Assembly to discuss the best means of raising recruits for the fleet. Speaking before the body, Penn and Venables pointed out the advantages that would result from the successful completion of their mission, and demanded no less than 4,000 recruits from Barbados. Venables asked that he be given a list of all the freemen on the island, so as to avoid the enlistment of servants.[32]

The islanders were thrown into an uproar by the speeches. When the Assembly went back to its chamber, Speaker Modyford urged the acceptance of the request, arguing that the colony should be thankful that servants were not to be recruited; but no one in the Assembly was inclined to listen to his opinion, particularly as he had helped to bring the expedition to Barbados. "I found such a willfully embittered party," Modyford wrote, "that instead of debating calmly they fell clamouring against the quartering of soldiers in their houses, their rudeness and misdemeanours, and would come to no conclusion but this, let them beat up drums and take their course, we will not assist them."[33]

With householders unwilling to give any assistance, recruiting agents from the fleet took volunteers and impressed individuals without too many questions being asked. Free laborers, debtors, and bonded servants were all accepted for service and hurried on board. Modyford reported that "though the commissioners sent out strict orders to their officers not to list servants, yet their indentures not being writt in their forehead, they were by some ignorantly and by others wilfully received; and when once they were got into the huddle there was no finding them."[34] Lt. Col. Francis Barrington, a recruiting officer, wrote home that "the doing of this hath much injured the poor people even to their undoing and prejudiced many of the rich, some losing ten servants, some fifteen, some more some less, none escaping us; therefore most men will conjecture, hearing of it, that we dealt very severely with our countrymen."[35]

The outrage of planters was further fueled when the General Assembly was asked to hand over 2,000 muskets from the regiments of the militia. Surrounded in their households by a dangerous mix of slaves and servants, planters were being asked to give up some of their weapons. Fearful for their safety, they resisted the order, but Penn and Venables commanded their officers to collect the arms by force—"in which proceedings misdemeanours were committed by divers of the soldiers."[36] One officer recalled that "we took [firearms] where we could find them, without giving any satisfaction to the owners."[37]

Adding insult to injury, Barbados was ordered to provide financial support for Cromwell's Western Design. In March 1655 an astonished General Assembly was informed by General Venables that he had been instructed to appropriate the excise duty for the carrying on of the expedition. Planters viewed the action as the imposition of a special tax on the island and, as such, a violation of the articles of surrender that guaranteed that no tax would be imposed without their consent, but the order remained enforced.[38]

If planters had any doubts about their colonial status before the arrival of the fleet, by the time it departed there could be no doubt left in their minds. The experience put them in a dark and ugly mood. Writing to his brother in England, Modyford related that there was "such a strange cursing and railing at these men after they were gone, that it would have troubled your ears to have heard it."[39]

Frustrated by their treatment, planters were soon involved in yet another controversy with the mother country. To their chagrin, in the fall of 1655 they learned that the governor was not the only public official whom they had to accept as a commissioned officer from England. Slowly but undeniably a pattern emerged in the second half of the seventeenth century in which more and more of the island's public officials were selected in England. This trend was not part of a grand imperial strategy to build a colonial bureaucracy through which England could more efficiently govern the colony. In the case of lesser public officials (provost marshal, secretary, clerk of the markets), the initiative to have them appointed in the metropolis usually began with the petition of an eager place seeker to the Lord Protector or, later, the king. It was the lucrative fees that were associated with these offices that prompted their application. When it saw fit to grant one of these requests, the imperial authority was motivated by the desire to reward a loyal political supporter. In the daily execution of these lesser posts there was no exchange of information between the metropolis and these officeholders, nor did the latter operate under specific orders from the central government. Yet, in effect, their appointment created a group of commissioned officers in the island's government, many of whom had no proprietary interest in Barbados at the time of their selection, and whose loyalties were to their English patrons.

Neither the island's governor nor the plantocracy was pleased by the emergence of patent officers, as these English appointees came to be called. Governors complained that they should have the right to select local officeholders. The patronage, they argued, would give them political leverage in the colony, enabling them to build a responsive faction in the planter community.[40] The plantocracy recognized that the commissioning of Barbadian public officials in England would tend to limit their chances of holding office, because close ties to the source of patronage would determine who acquired the position. The plantocracy also deeply distrusted patent officers, seeing them as men whose eagerness to improve their material circumstances would lead to excesses in office.[41]

The disagreement between the metropolis and the colony over patent officers began in the mid-1650s. While the proprietors appointed the island's provost marshal and secretary in the 1630s, during the period of independence in the 1640s all lesser public officials were selected by the governor and the Council, with the Assembly setting down in minute detail what fees these men could charge in the performance of their duties.[42] Following the latter practice, in 1654 Searle appointed William Povey the island's provost marshal and Thomas Noel its secretary. They were the brothers of two English merchants, Thomas Povey and Martin Noel, whose political stars were on the rise in England during the interregnum. Actually, the latter Povey was the secretary of Martin; together they played a significant role in English colonial affairs for much of the 1650s.[43] Through his connections, Martin was able to get William and his brother, Thomas Noel, commissions from the Lord Protector for their offices in Barbados.[44] The governor was not at all happy that the island's provost marshal and secretary were no longer under his immediate control.[45] The plantocracy was equally displeased that these men were now appointees of the imperial authority. Even William Hilliard, the strategist for the reduction of the island in 1652, and James Drax, who gained politically by the parliamentary takeover of the island, shared the general displeasure over the rise of patent offices on the island. Thomas Povey told his brother that he found Drax and Hilliard extremely unwilling that you "should be confirmed here by patent . . . otherwise they are very hearty in your behalf."[46]

To mollify Barbadians, Povey assured them that the appointments would not be taken as a precedent in England, presumably meaning that in the future the governor would decide who sat in these offices. He warned planters, however, that they should not be "so tender and sensitive of every act of power which your prince shall think fit to exercise amongst you." Some extraordinary things, he predicted, "they [princes] may and will sometimes do to remember us of their sovereignty and love." In any case, he told them, princes "should not be told of what they cannot nor ought not to do, especially when they put forth their power rarely and with caution."[47] In the face of the protest of Barbadians, Noel and Povey took their offices under a commission from the mother country.

It was fortunate for Searle that the dispute over patent offices brought him into contact with Thomas Povey and Martin Noel. In England, Colonel Modyford, through friends, was urging the recall of the governor, and the promotion of himself in his stead. It would have gone hard for Searle, Povey predicted, but for the steady and powerful backing of Noel. Indeed, realizing the strength of the governor's support, Modyford decided to make a momentary show of reconciliation, hoping that a more favorable opportunity might allow him to press successfully his case.[48]

The occasion presented itself at the death of Oliver Cromwell on September 3, 1658. Cromwell's son Richard became the new Lord Protector, but Richard was not up to the task of guiding England through the murky political waters that the English Civil War had left in its wake. In April 1659, Richard dissolved the Parliament that he had summoned, and a few weeks later he resigned his office. The army leaders thereupon convened the remnants of the Long Parliament.

On April 30, 1659, Povey and Noel sent a joint letter to Searle telling him about the unsteady political situation in England. They advised the governor to take no formal notice of the changes at home until an official notification was sent to him. "Nevertheless it may be prudent in you to behave yourself as one who are now to be accountable to a commonwealth." The personal influence of Povey and Noel had greatly decreased by the change of government, but they promised to do their utmost to keep Searle in office.[49]

A commission confirming all officers in their appointment was sent to the colony by the Council of State on June 6. Writing two days later, Povey warned Searle of the precarious nature of his position; many rivals for the governorship would now appear. Searle would have to tread warily both in his relations with Whitehall and at Barbados, "it being a tender thing for one in your place to keep a reverence, and enforce an authority without a commission." Povey advised the governor to leave the Parliament alone as much as possible—it was busy and therefore irritable.[50]

Searle had to be cautious in these uncertain times, but Barbadians saw the instability as an excellent occasion to renew their demands for local independence and freedom from imperial restraint. On December 11 the islanders sent a petition to Parliament. In addition to the usual demands for free trade, the petitioners asked "that we may have a confirmation of liberty here (by a law or your commission) for the representative body of the people to choose a governor out of the freeholders of this land and one out of every parish to be his assistants and join with him the execution of government." They also requested that they have complete control over the appointment of all officials on the island, that all legal profits should be at the disposal of the Barbadian government for defraying public expenses, and that they might have the privilege (already accorded to New England and Jamaica) of coining their own money. In fact, the status demanded was inferior to complete independence in little else but name.[51]

Owing to the confused political situation that existed in England in late 1659 and early 1660, the freedom to manage their own affairs, as they had done in the 1640s, seemed to be once again within the grasp of planters. On April 24, 1660, Colonel Modyford was appointed governor of Barbados. The commission he received was Parliament's answer to the planters' petition. Although the right of electing their own governor was not granted, a compromise was effected in that for the first time since the early days of settlement the nominee was a planter of the island. Hitherto the home government had always appointed strangers, men who had little or no knowledge of Barbadian interests. As governor, Modyford stated in his inaugural speech to the General Assembly, "I am the first of this order, planter-governor, and there is hope I shall not be the last . . . but that every one of them that stand before me this day may according to their merits have a turn and share at the helm."[52] The island's Council was to be appointed by local elections, each parish selecting one member. Thus, with one of themselves as governor, and with the Council and the Assembly as their own elected representatives, the plantocracy possessed a very considerable measure of self-government, but the newly acquired privileges were destined never to be put into execution.

Toward the end of May 1660 word reached the colony of the great preparations in England for the king's return. The news of the Restoration must have deflated the spirits of planters, as it meant that Modyford's commission would probably be recalled. Nevertheless, on July 16, 1660, Modyford took over the governorship from Daniel Searle and made a bid to continue in power by becoming a Royalist. Once in office, he proclaimed Charles II king of England and her dependencies. Through the influence of his cousin, General Monk—who helped to engineer the return of the monarch—Modyford received a full pardon for his betrayal of the Royalist cause at Exeter, Barbados, and his subsequent services to the Commonwealth.[53]

As Modyford was acquiring a pardon for himself, the king directed Francis Lord Willoughby to undertake the government of the several islands of the province of Carliola (including Barbados) in accordance with the powers granted by letters patent to the earl of Carlisle. The revitalization of the proprietary rule raised a host of unsettling problems for Barbadian householders. According to past agreements with the landlords, the rents of tenants were uncertain, and any duty or tax that the proprietor thought fit he could impose. Equally worrisome, no one—small farmer or large sugar producer—had a clear and certain title to his land. It is therefore not surprising that Barbadians protested vehemently against the resuscitation of the proprietary patent. They were supported in their actions by Modyford (who saw in Willoughby a rival for the governorship) and by London investors in the colony. After much discussion, in which the earl of Kinnoul, heir to the second earl of Carlisle, and the creditors of the first earl (who had never been paid off), put in their claims against the proprietorship, the attorney general of England ruled that the Carlisle patent was invalid.[54]

Needing to stay in England while the fate of the island was determined, Willoughby exercised his authority under the king's provisional order and appointed Col. Humphrey Walrond as his deputy. Anticipating that the Carlisle patent would be withdrawn, Willoughby suggested that he should continue as the governor of Barbados under the king, who should himself assume the proprietorship of the islands and receive half of the profits therefrom, the other half accruing to Willoughby. The agent of Barbadian planters in London made a somewhat similar proposal, which involved royal control of the colony and the payment of an export tax on the produce of the island. The colonists would later complain that their agent was not authorized by them to suggest an export tax, but their disavowal went unheeded.[55]

The recommendations appealed to the king, who needed some revenue and wanted to find a solution to the long dispute over the proprietorship of Barbados. In June 1663 he appointed Willoughby governor of the colony for seven years, the remaining period of his lease of the islands from Lord Carlisle. Willoughby was to receive half of the profits and to defray all the expenses of the government. The other half of the profits was to be used to pay an annual sum of £500 for two lives to the earl of Marlborough, to pay an annual sum of £500 to Lord Kinnoul in perpetuity, and to pay off the creditors of the first earl of Carlisle. When the creditors were fully paid, Lord Kinnoul's annuity was to be doubled, and the remainder was to revert to the king. In fact, the creditors never received anything. Owing to the heavy war expenditure, there were no profits for many years, and later their claims were ignored.[56]

In August 1663, Lord Willoughby returned to Barbados after an absence of eleven years. He immediately summoned the Assembly and urged it to grant the king a tax of 4.5 percent on the value of all commodities grown and exported from the island. In exchange, the governor promised to give planters confirmation of their lands in spite of doubtful titles and to abolish all dues and rents formerly payable to the proprietor. After much opposition, the Assembly accepted the compromise.[57]

The tax so imposed remained in force until 1838. It was a constant source of irritation in Anglo-Barbadian relations. The mother country maintained that the tax was no more than a payment for the abolition of proprietary dues and the confirmation of land titles that were admittedly defective. Barbadians claimed that the tax was intended to meet the expenses of the island's government. Even though the terms of the act that embodied the compromises would seem to confirm the planters' argument, there was little the colonists could do to enforce compliance by the metropolis. In effect, the 4.5 percent duty became a colonial tribute that served as a reminder to Barbadians of their subordination to England. It was seldom spent to local advantage; and additional taxes had to be levied to cover the public expenditures of the island, as quickly became evident in the series of imperial wars that broke out in the West Indies in the second half of the seventeenth century.[58]

On April 20, 1665, the Dutch under Admiral Michel de Ruyter, with fourteen ships, arrived at Carlisle Bay and attacked English ships at anchor there and the fort on land. The battle began at nine o'clock in the morning and continued until four o'clock in the afternoon, when de Ruyter broke off the attack and sailed away.[59]

The immediate danger was over, but Barbados had to prepare itself for the possibility of another assault. Reluctantly, Willoughby called a meeting of the General Assembly and requested that further taxes be imposed to cover the estimated cost of defense. His demand was opposed by the Assembly, the members arguing that the expenditures should be met from the proceeds of the 4.5 percent duty. Seeing that it was not going to give him what he wanted, Willoughby prorogued the lower house. Early in 1666, however—when, in addition to the danger of a Dutch attack war with France appeared to be imminent—Willoughby summoned another Assembly and reiterated his request for new taxes. The governor was once again told that the export duty should be used for that purpose, but he retorted that royal instructions were very definitely against this policy. After much wrangling, the Assembly voted the necessary supplies in the interest of public safety.[60]

In January 1666 the English island of St. Kitts fell to the French. In July, by order of the king, Willoughby was to lead an expedition from Barbados to recapture the colony. With a fleet of 8 ships and about 1,000 soldiers, he sailed from Barbados on July 18, 1666. A few days later, a violent hurricane suddenly struck the fleet off the coast of Monserrat. Lord Willoughby's ship was lost in the storm, and he was never seen again. Before his departure from the island, he had appointed his nephew, William Willoughby, lieutenant governor of Barbados.[61]

William did his utmost to awaken the mother country to the dangerous situation that existed in Barbados in regard to the shortage of recruits for the colony's militia. With the fall of St. Kitts a recent occurrence, shortly after his taking office, he sent an anxious report to Charles II in which he described the state of the Barbadian militia and its inability to defend the colony. Willoughby informed his sovereign that there were 7,000 men on Barbados able to bear arms but that only 2,000 were those "whose interest and honour" would cause them to make a resolute defense of the island. The remainder consisted of three sorts of men, and none of them could be relied upon in an emergency. The first group were small planters who were so "impoverished and disheartened that the welfare of the colony was but little esteemed by them." The second group he called freemen, or those "having no estate of land," getting their livings by daily labor. These men had undergone harsh servitude in Barbados and were without the means to improve their condition. In Willoughby's opinion, they were willing to "stoop to any alteration" to reverse their circumstances, and hence they could not be trusted as militiamen. Equally untrustworthy were indentured servants. They were accustomed to "slender allowance and hardship" in Barbados, and the governor discerned a common desire among them to serve "new masters." Thus, Willoughby depicted the island's military predicament and "what little dependence may be had on many of its inhabitants." The greater number being without "interest or hopes of benefit" in the society, it was much doubted whether they would expose themselves to danger for its preservation.[62]

A striking omission from Willoughby's letter was his failure to comment on the serviceability of slaves, who represented slightly more than half of the population in the late 1660s. Presumably, he felt that their position in the society made them so obviously useless as militiamen that he need not trouble his sovereign with a comment. Yet, by a proclamation apparently issued before the report was written, it was ordered that "two suitable Negroes" should accompany each trooper on alarms.[63] This directive did not formally induct blacks into the militia, which was restricted to whites, but it certainly placed some blacks in a position to fight for the island if the need should arise. No doubt the threat of invasion during the Second Anglo-Dutch War encouraged planters to make up some of the colony's military deficiencies through the employment of a select group of slaves.

Disregarding the fact that he did not remark on over half of the island's population, the governor's letter to the king was still an astute characterization of the state of morale in the militia and the island's need of English assistance. It helps confirm the view that social and economic conditions in Barbados were undermining the colony's ability to defend itself.

In January 1667 the king appointed William Lord Willoughby (brother and heir of Francis and father of William) governor of Barbados. He arrived on the island in April 1667, and he rapidly learned about the precarious military state that existed there. In the fall of 1667, he warned the English Privy Council that the "great want of servants" would bring the safety of Barbados into question. "For though there be no enemy abroad," he reminded them, "the keeping of slaves in subjection must still be provided for."[64] At the same time, the Barbadian Assembly directed a petition to the crown, asking that 1,000 or 2,000 English servants be sent to the colony.[65] An English commentator on their supplication cut to the heart of the planters' dilemma. Any persons who were sent over, he observed, "could not be judged constant to the purpose" of the plantocracy who wanted them for the island's defense, but "will as all opportunity presents, as others their predecessors void of propriety have done at the expiration of their servitude, seek their fortunes abroad. This will put the island in the same condition it's now in, and the king put to the same trouble for supply."[66]

Willoughby told the crown that Barbados would never be capable of defeating a foreign enemy until "some way be found to give a comfortable living to the meaner sort." He recommended that 10 acres in every 100 be taken from the rich and given to the poor, but he doubted whether the plantocracy would be willing to initiate policies that would improve the material life of poor whites on the island.[67]

It was during Willoughby's administration that a serious attempt was made to reform the island's slave plantation system in order to provide reliable recruits for the island's militia. This new wave of concern about the militia was prompted by reports in 1671 from absentee planters in London that England might soon be at war with France, so that the islanders should be "thinking of how to defend themselves." Their advice to the colonists shows how well it was understood that economic conditions were undermining the colony's ability to protect itself. Being informed that "2,000 people are gone off Barbados this last year [1670] and more are still going," they suggested that "no man possessed of land in Barbados be capable of purchasing any more, which will uphold the number of freeholders." In a later letter, it appears that the absentee planters refined their thoughts on how the government should go about the business of halting further land consolidation. They now proposed that "no one possessed of twenty-five acres of land shall be capable of buying, renting or receiving more unless by descent, forfeiting all lands so purchased to the first man that had not twenty-five acres that enters actions for it in the court of the precinct where it lies." If this land reform were not adopted, they predicted that all the acreage on the island would "fall into the hands of a few, and they will be lost for want of enough interested men to defend the place." They also advised that Negroes and other servants be clothed by the "manufacture of Barbados" instead of by imports, "which would find employment for many of the poor, who go off because they know not how to subsist." Finally, they recommended that restrictions be placed on the employment of Negroes in skilled occupations, the practice that many thought had caused white tradesmen to leave the island.[68] In short, to improve Barbados's defensive capabilities, they suggested that employment for free labor be found.

Inspired by the proposals of the absentee planters, in 1671 the Barbadian General Assembly passed an "Act to Prevent Depopulation." The "safety and prosperity of this island," the preamble to the statute began, "does chiefly depend in the number and strength of Christian inhabitants." It was they "who on all occasions will doubtless prove themselves most serviceable and ready both in the resisting of foreign invasions and also putting a stop to intestine insurrections." Desiring that an adequate number of whites "may not at any time hereafter be found wanting," the Assembly was moved to make some reforms in the society.[69]

The bill fell far short of the recommendations of the London planters. On the matter of controlling land consolidation, the Assembly chose to make it illegal for those who acquired land to pull down or cause to decay the "manor house" situated on the property, unless they built a cottage or house elsewhere on the land. This law was certainly a far cry from that suggested by the absentee planters, which would have stopped land consolidation except by inheritance. The provision was also ambiguous in that it was unclear whether the humble dwelling of a small farmer was to be treated as a "manor house." More closely related to the needs of the poor, the General Assembly sought to encourage big landowners to lease their land to small planters by allowing them for militia purposes to rate two tenants who had leases of at least three years on plots of two acres or more as the equivalent of three freemen or servants. The effect of this clause was to create a class of military tenants who held their lands primarily because of their service in the militia. The Assembly also ordered plantation owners to hire one qualified white for every slave working at a trade. In that same regard, trying to improve the living standards of poor whites in the seaport towns, whom it was thought might have a more comfortable existence were it not for the "superabundant number of Negroes" who could in "no way lend to the safety of the place," it was stipulated that masters of slaves in the towns had to keep one Christian man for every Negro bondsman.[70]

Notably absent from the law was any provision to encourage the development of a textile industry or any regulation that required planters to clothe their servants and slaves in garments manufactured in the colony. In a letter to the absentee planters, a committee of the Barbadian Assembly mentioned that a separate piece of legislation was soon to be passed that would promote the making and wearing of apparel made on the island, but there is no evidence that such a statute was ever enacted.[71]

The bill had no tangible effect on the emigration rate, and there is no reason to suspect that it improved the fighting ability of the trainband. But even this ineffectual legislation soon proved a greater burden than the plantocracy cared to shoulder. The whole act was repealed in 1688, the General Assembly falsely maintaining that some of the clauses were covered in other legislation, that some were unnecessary, and that others were contradictory.[72] So ended the most concerted effort to reform the Barbadian plantation system in the seventeenth century. In refusing to improve the economic situation for free labor in Barbados, the plantocracy tacitly accepted the colonial relationship.

In 1673, William Willoughby died, and Sir Jonathan Atkins succeed him as governor. Atkins was extremely sympathetic to planters' concerns, which frequently brought him into trouble with the mother country and finally resulted in his recall. For example, he supported the Barbadian claim that lesser public officials should be selected in Barbados rather than in England. The issue had arisen again in the winter of 1676 when Charles II chose to make the clerk of the market a patent officer. The island's General Assembly angrily wrote the monarch that "it is prejudicial to government to have officers nominated" in the metropolis, questioning whether a man commissioned in England would faithfully discharge his duties in the colony.[73] Acting as the king's agent in this matter, Henry Coventry told the islanders that he hoped the General Assembly was "not so insolent as to declare to the king who was fit or not fit to serve him." He asked planters if they had anything to "show whereby the king had deprived himself of bestowing this place." He doubted, however, that the General Assembly could take it upon itself to "erect any offices with power to take fees of the king's subjects and by their power exclude the king from disposing" of the office. In concluding his letter, he informed planters that the king was resolved to continue his grant, but if the colonists wanted to challenge the king's judgment he advised them to give such "reasons as may satisfy His Majesty in council or his judges in Westminster Hall, for the last determination will not be in Barbados."[74]

In that final phrase, "the last determination will not be in Barbados," Coventry laid bare a basic fact about Anglo-Barbadian relations. In any dispute between the mother country and the colony, the ruling powers in England would decide the issue. It mattered not how contrary their opinions and interests might be; planters had to be responsive to the dictates of the mother country, though they grumbled and were inclined to disobey. Social and military problems that were an outgrowth of the island's plantation economy made it impossible for them to resist imperial authority. In fact, as racial polarization became a factor in the island's maturing plantation system in the second half of the seventeenth century, their dependency on England was further solidified.

CHAPTER 9

Race, Racism, and the Imperial System

English settlers came to Barbados making assumptions about the mental and physical abilities of the races, and from the beginning they referred to Africans as Negroes and blacks, but religion had more social relevance than race in seventeenth-century England, and planters sought to establish slavery on theological grounds.[1] By the end of the century, however, the religious justification for the enslavement of Africans had all but disappeared, having been replaced by notions that blacks were naturally inferior to whites and thus should be their slaves. These ideas legitimated the inequities that existed in most Barbadian households, and in so doing they contributed to the island's continued dependency on England.

The most useful source of information on the racial composition of the colony's households is the census of 1679.[2] It is an alphabetical list, parish by parish, of the 2,369 Barbadian property holders, with the number of acres, servants, and slaves belonging to each; there is also a list of the 405 households in Bridgetown showing which were married and the number of children, servants, and slaves for each. For purposes of analysis, it has been assumed that every individual who was reported on in the census was a householder. In cases where an individual owned property in more than one parish, his name appears more than once in the census. Rather than combine his holdings, which could not be done with precision because of the difficulty of identifying all persons reported on, each entry is treated as a separate household.

Table 9.1 presents a summary of slaveholding patterns in Barbados drawn from the census. Households have been arranged by acreage held: those with no land, those with 1 to 9 acres, those with 10 to 99 acres, and those with 100 or more acres. While this division may appear to be arbitrary, the acreage of a plantation had a direct bearing on the social and economic relations within households, and in those terms the division will have meaning.

About 4 percent of the island's slave population lived in Bridgetown, the island's capital and chief port of call. These urban blacks resided in households in which the number of whites was roughly equal to the number of blacks.[3] The typical household consisted of a husband and wife, 1 white child, 1 servant, and 4 slaves. None of the slaveholders in Bridgetown owned more than 20 Negroes.

Bridgetown blacks were not agricultural laborers. In some households they were primarily employed as domestic servants by masters who preferred to live in the more cosmopolitan atmosphere of Bridgetown than in the isolation of their plantations. William Bate, for example, had a 125-acre sugar estate in St. Michael on which over 100 slaves worked, but he kept a house in the city with 1 slave to tend to his needs.[4] Those with relatively large numbers of slaves (John Hassel with 20 or David de Mercado with 11) were merchants.[5] Their blacks worked in the city's warehouses, shops, and taverns.

Because slaveowners in Bridgetown were not interested in plantation laborers who could be organized into gangs but wanted slaves who could more or less be left to work on their own, blacks in the city were generally more familiar with European ways and spoke English. One planter estimated in 1668 that there were "thousands of slaves that speak English" in the colony. He thought they were "either born there [Barbados] or brought young into the country."[6] Many of these Creole slaves, as planters called them, lived in Bridgetown, but they could also be found in the countryside, particularly in the households of small planters.

TABLE 9.1

Slaveholding Patterns (1679/80)

Parish	0 Acres				1–9 Acres				10–99 Acres				100+ Acres			
	No. owners	*No. slaves*	*% of slaves (parish)*	*Mean no. slaves*	*No. owners*	*No. slaves*	*% of slaves (parish)*	*Mean no. slaves*	*No. owners*	*No. slaves*	*% of slaves (parish)*	*Mean no. slaves*	*No. owners*	*No. slaves*	*% of slaves (parish)*	*Mean no. slaves*
St. Peter	81	419	13	5	12	49	2	4	90	1,190	38	13	21	1,503	48	72
St. Andrew	28	86	4	3	8	16	1	2	43	633	30	15	20	1,395	65	70
St. Thomas	6	31	1	5	31	99	3	3	74	982	30	13	24	2,147	66	89
St. John	14	82	3	6	14	45	1	3	49	860	27	18	24	2,204	69	92
St. Lucy	17	31	2	2	61	122	6	2	145	1,032	52	7	8	788	40	99
Christ Church	23	94	2	4	84	250	5	3	154	1,503	32	10	29	2,858	61	99
St. Philip	5	75	2	15	84	204	5	3	184	1,978	44	11	28	2,224	50	79
St. James	13	46	2	4	23	76	3	3	57	727	24	13	21	2,163	72	108
St. Joseph	0	0	0	0	12	47	2	4	31	550	26	18	15	1,533	72	102
St. George	8	30	1	4	8	51	1	6	52	932	22	18	29	3,299	77	114
St. Michael	7	27	1	4	55	266	7	5	100	1,478	40	15	20	1,919	52	96
Total	195	921	2.8	4	397	1,225	3.2	3	979	11,865	33	12	239	22,033	61	92

Source: P.R.O., C.O. 1/44/149–241.

Roughly 3 percent of all Barbadian slaves were owned by individuals who held no land. The overwhelming majority of these slaves (82 percent) lived in households in which there were no servants (see Table 9.2). The mean number of blacks held by landless slaveholders was 4. If the size of white families in Bridgetown is representative of planter familial patterns on the island, blacks in the households of landless slaveowners who held no servants outnumbered whites 4 to 3. In those households in which both slaves and servants were to be found the number of whites and blacks were roughly equal—4.5 whites to 4.6 blacks (see Table 9.3).

Most landless slaveholders were in all likelihood tenants, farming relatively small tracts of land. The types of crops they cultivated had a direct bearing on the slave's existence. If they attempted to grow sugar, they could plant only small amounts on their rented lands, so that they probably raised provisions for local consumption, or tobacco or cotton for the world market. The small amounts of sugar these slaveholders farmed meant that their blacks were not subjected to the eighteen-hour workdays that were commonplace on large sugar estates during the harvest season; nor were their Negroes accustomed to laboring in plantation gangs. The more acculturated blacks frequently worked without their master's immediate supervision. Newly transported Africans either worked under their owner's watchful eye or were assigned to one of the creolized bondsmen in the household. In either case, the fact that they labored in close proximity to someone fluent in English helped to quicken their acculturation to plantation life.[7]

Although these landless slaveholders were not the abject poor of Barbados, their purchase of African labor represented a sizable investment for them. In many instances, they did not buy their bondsmen outright but went into substantial debt to expand the size of the labor force in their household, and what they feared most was that their blacks would die before they could make good on their investment. Edward Littleton, author of an influential pamphlet on the difficulties of managing a West Indian plantation, wrote that "planters were in hard conditions when a pestilence decimated their slaves, especially if he be still indebted for them. He must have more Negroes, or his works must stand, and he must be ruin'd at once."[8] Littleton probably had large estate owners in mind when he wrote those lines,

TABLE 9.2

Slaves in Households without Servants and No Land

Parish	(1) *No. landless slaveholders*	(2) *No. landless slaveholders w/o servants*	(3) *% landless slaveholders w/o servants*	(4) *No. slaves held by landless slaveholders w/o servants*	(5) *(4) as % of total no. slaves in parish*	(6) *Mean no. slaves in household of (2)*
St. Peter	81	50	61	210	50	4
St. Andrew	28	28	100	86	100	4
St. Thomas	6	5	83	27	87	5
St. John	14	14	100	82	100	6
St. Lucy	17	14	82	31	87	2
Christ Church	23	19	83	94	63	3
St. Philip	5	5	100	75	100	15
St. James	13	10	77	36	78	4
St. Joseph	0	0	0	0	0	0
St. George	8	8	63	30	67	4
St. Michael	7	7	86	27	89	4

Source: P.R.O., C.O. 1/44/149–241.

TABLE 9.3
Mean Number of Slaves and Servants in Households with No Land

Parish	*Slaves*	*Servants*
St. Peter	6.7	2.3
St. Lucy	1.3	1.0
Christ Church	8.7	1.7
St. James	3.3	1.3
St. George	3.3	1.3
Average	4.6	1.5

Source: P.R.O., C.O. 1/44/149–241. (White family size 3).

but his words were equally true for small slaveholders. Within the limits of profit, the latter attempted to see that their bondsmen received the minimum required to sustain life. Emulating large estates, they built thatched huts to serve as quarters for their slaves. Some may have fed their bondsmen from their own pot; others portioned out provisions, letting blacks cook for themselves. Among big planters in Barbados there was a general consensus that Negroes needed only enough clothing to cover their nudity, and it is unlikely that lesser slaveowners gave their bondsmen much above that allowance.[9]

Beyond financial concerns about the physical well-being of their Negroes, as with slaves in Bridgetown, the comparatively small number of blacks held by these planters made frequent contact between masters and slaves possible. The proximity permitted emotional attachments to develop that tended to blunt some of the sharper edges of slavery. While not frequently done, it was not unheard of for a slaveholder out of genuine affection to manumit a black upon his death. Philip Bell in his will ordered that after his wife's death Arbella was to be set free with Mango, a "Negro man . . . who used to run along with me."[10] In 1657, Hanna Adcock freed her slave Sarah.[11] In 1679 the master of Asha stipulated that she was to "serve for the space of twelve months after my decease," and then she was to be manumitted.[12]

A few landless slaveholders had as many as 20 slaves. Henry James and John Swimsteed of the parish of St. Peter both had 40 slaves even though they possessed no land.[13] They were in the business of hiring out Negroes to other planters; the latter was responsible for feeding and housing the bondsmen for the duration of the contract.[14] This practice undoubtedly began in the 1640s when men who were unable to purchase arable land invested in slaves instead. There must have been occasions when a small planter rented a slave or two for a few days, but usually they were hired by large estate owners. For big planters the renting of slaves had the advantage of keeping their own bondsmen from being overworked while making it unnecessary for them to purchase more Negroes than they could profitably employ the year round. Commonly the rentals were put to labor at the most physically demanding tasks on the plantation, such as digging cane holes, thus sparing the planter's regular bondsmen from work that might have proved physically debilitating. Since the hiring of blacks was much more expensive than the outright buying and maintaining of bondsmen, planters tried to avoid the practice as much as possible.[15]

The life of a hired slave was markedly different from that experienced by bondsmen of landless slaveowners. While their masters certainly had an economic interest in seeing that they were not severely mistreated, the fact that they often did strenuous labor meant that their mortality rate was higher than was normal for the rest of the island's population. One might also expect that the number of newly transported Africans among hired Negroes was higher than was the norm for blacks belonging to households of landless slaveholders. Furthermore, since these slaves constantly moved from one plantation to another, it was difficult for them to establish close ties with their master or with any white, for that matter. In short, blacks owned by men who were in the business of renting Negroes were less acculturated and more exposed to the rigors of slavery than was true for most Negroes in the possession of landless slaveowners.

Approximately 4 percent of Barbadian slaves lived on a small farm of less than 10 acres. The mean number of bondsmen on these tiny plantations was 3. Eighty-three percent of these blacks belonged to households in which there were no servants, though the number of Negroes and whites was roughly equal (see Table 9.4). About 12 percent of those landholders who possessed between 1 and 9 acres and owned slaves also maintained servants in their households. The mean number of slaves was 3.8, and there were about 2.6 servants on these

TABLE 9.4

Slaves in Households without Servants and 1–9 Acres

Parish	*(1)* *No. slaveholders 1–9 acres*	*(2)* *No. slaveholders w/o servants*	*(3)* *% slaveholders w/o servants*	*(4)* *No. slaves held by slaveholders w/o servants*	*(5)* *(4) as % of total no. slaves in parish*	*(6)* *Mean no. slaves in household of (2)*
St. Peter	12	9	75	37	76	4
St. Andrew	8	8	100	16	100	2
St. Thomas	31	30	97	92	93	3
St. John	14	13	93	40	89	3
St. Lucy	61	58	95	122	90	2
Christ Church	84	76	90	214	87	3
St. Philip	84	84	100	204	100	2
St. James	23	20	87	65	86	3
St. Joseph	12	10	83	23	49	2
St. George	8	6	75	47	92	8
St. Michael	55	38	69	138	52	4

Source: P.R.O., C.O. 1/44/149–241.

TABLE 9.5
Mean Number of Slaves and Servants in Households with 1–9 Acres

Parish	*Slaves*	*Servants*
St. Peter	4.0	1.6
St. John	5.0	1.0
St. Lucy	4.0	4.6
Christ Church	4.5	1.5
St. James	3.6	1.0
St. Joseph	2.0	7.0
St. George	2.0	1.5
St. Michael	5.3	2.6
Average	3.8	2.6

Source: P.R.O., C.O. 1/44/149–241.

farms (see Table 9.5). Since these masters of servants cultivated about as much acreage as landless slaveowners who rented their farms, it can be assumed that these bondsmen worked and lived in conditions that were similar to those for blacks owned by planters who possessed no land.

Thirty-three percent of the Barbadian slave population belonged to what might properly be called the Barbadian middle class. These planters owned somewhere between 10 and 99 acres, and held about 14 slaves. Clearly, as the broad differences in land would suggest, not all these planters fit well into this category. William Butler of Christ Church, who possessed 10 acres and 1 slave, for example, could not have been much wealthier than Thomas Johnston, who lived in St. Andrew with 8 slaves and no land.[16] Certainly, Andrew Alick of St. James was better off than both of them with his 96 acres and 70 slaves; indeed, he might well be considered a big planter.[17] Gabriel Deane of St. George is perhaps more representative of the average middling planter. He owned 22 acres and 14 slaves.[18] The one thing that William Butler had in common with Andrew Alick and Gabriel Deane, which separated him from Thomas Johnston, was that his possession of 10 acres qualified him to vote in Barbadian elections, so it seems proper to see him as part of the middling sort, even though in terms of actual wealth it might be better to think of him as a small farmer.

Approximately 52 percent of the slaves owned by middling planters lived in households in which there were no servants and a mean number of 10 slaves (see Table 9.6). The other 48 percent belonged to planters who held 2 servants and 24 slaves (see Table 9.7). In the household of the Barbadian middle class, whites tended to be outnumbered 3 to 1.

Slaves who were the property of lower-middle-class planters, the William Butlers of Barbados, were accustomed to social conditions and work routines that were not much different from those of Negroes who were owned by relatively prosperous landless freemen or small proprietors. They did not labor in gangs, and only a small portion of their time was devoted to sugar cultivation. Slaves possessed by the upper middle class, the Andrew Alicks, however, lived more closely to the life-style that was common for the majority of the island's Negroes.

About 61 percent of all black bondsmen in the colony resided on a plantation of 100 acres, which had a mean number of 98 slaves. There were, of course, much larger estates. Christopher Codrington had 618 acres and 300 slaves.[19] Samuel Newton owned 581 acres and 260 slaves.[20] In contrast to the farms of small and lower middling planters, only 11 percent of these blacks were members of households in which there were no servants (see Table 9.8). Eighty-nine percent stayed on estates in which there were 104.4 slaves to 5.1 servants (see Table 9.9). On these plantations, whites were outnumbered 13 to 1.

Unlike the urbanized blacks of Bridgetown who were exposed to all kinds of currents in the island's chief port of call, Negroes on large sugar estates lived in comparative isolation. Except for Saturday afternoons and Sundays, they spent much of their day on their master's plantation. Because of the size of their owner's household, it was difficult for these blacks to develop close personal relationships with their masters. It was a rare instance when they were invited into the big house. By and large, they worked in gangs, and they were governed through a highly structured bureaucracy that extended from their master to white overseers, culminating in the black drivers who labored in the fields with them. Under these circumstances, slavery was more impersonal, and the acculturation of "new Negroes" occurred at a slower pace than in the households of small planters.[21]

The differential rates by which Africans in small and large households adjusted to plantation culture should not obscure the basic fact

TABLE 9.6

Slaves in Households without Servants and 10–99 Acres

Parish	(1) *No. slaveholders 10–99 acres*	(2) *No. slaveholders w/o servants*	(3) *% slaveholders w/o servants*	(4) *No. slaves held by slaveholders w/o servants*	(5) *(4) as % of total no. slaves in parish*	(6) *Mean no. slaves in household of (2)*
St. Peter	90	29	32	1,190	16	7
St. Andrew	43	38	88	491	76	13
St. Thomas	74	55	74	525	53	10
St. John	49	38	76	491	57	13
St. Lucy	145	116	80	588	57	5
Christ Church	154	138	90	1,193	79	9
St. Philip	184	154	84	1,357	69	9
St. James	57	43	75	331	46	8
St. Joseph	31	14	45	202	37	14
St. George	52	43	83	635	68	15
St. Michael	100	32	32	280	19	9

Source: P.R.O., C.O. 1/44/149–241.

TABLE 9.7
Mean Number of Slaves and Servants in Households with 10–99 Acres

Parish	*Slaves*	*Servants*
St. Peter	16.3	2.3
St. Andrew	28.4	2.0
St. Thomas	25.1	2.0
St. John	28.8	2.6
St. Lucy	15.3	2.0
Christ Church	19.3	1.6
St. Philip	20.7	1.4
St. James	28.3	1.9
St. Joseph	28.5	2.0
St. George	33.0	1.6
St. Michael	21.9	1.7
Average	24.0	2.0

Source: P.R.O., C.O. 1/44/149–241.

that all "new Negroes" had their world radically altered by the slave plantation system. The evolution of the slave family is suggestive of the enormous social and cultural changes they experienced in the colony.

The slave family was deeply influenced by black demographic patterns in Barbados. It is far from easy to assay slave demography in the seventeenth century, since planters kept no records of Negro births, marriages, and deaths; "but something can be attempted," writes Richard Dunn, "and all the evidence points towards a demographic catastrophe for the slaves."[22]

The simplest way to illustrate black mortality in Barbados is to compare the number of slaves imported during the course of the seventeenth century with the number of Negroes living in the colony at the close of the period. Between 1640 and 1700, according to Philip Curtin's careful calculations, about 131,000 Africans were carried to Barbados, at a rate of 2,400 per year; yet, by the end of the seventeenth century, only 40,000 were alive in the colony.[23] These figures are approximate, of course, but they prove without question that the Barbadian black population suffered an appalling natural decrease during the seventeenth century.[24]

TABLE 9.8

Slaves in Households without Servants and 100+ Acres

Parish	*(1) No. slaveholders 100+ acres*	*(2) No. slaveholders w/o servants*	*(3) % slaveholders w/o servants*	*(4) No. slaves held by slaveholders w/o servants*	*(5) (4) as % of total no. slaves in parish*	*(6) Mean no. slaves in household of (2)*
St. Peter	21	1	5	55	4	55
St. Andrew	20	8	40	458	33	57
St. Thomas	24	2	8	101	5	51
St. John	24	3	13	251	11	84
St. Lucy	8	1	13	51	6	51
Christ Church	29	23	26	282	10	47
St. Philip	28	5	18	281	13	56
St. James	21	2	10	135	6	68
St. Joseph	15	2	13	175	11	88
St. George	29	8	38	760	23	59
St. Michael	20	1	5	60	3	60

Source: P.R.O., C.O. 1/44/149–241.

TABLE 9.9
Mean Number of Slaves and Servants in Households with 100+ Acres

Parish	*Slaves*	*Servants*
St. Peter	72.4	5.0
St. Andrew	78.0	3.0
St. Thomas	102.9	3.8
St. John	118.7	3.8
St. Lucy	105.2	7.1
Christ Church	112.0	7.8
St. Philip	84.4	3.2
St. James	106.7	5.1
St. Joseph	135.0	6.2
St. George	120.0	4.7
St. Michael	113.1	6.4
Average	104.4	5.1

Source: P.R.O., C.O. 1/44/149–241.

The recent works of K. G. Davies and Richard Dunn on surviving West Indian slave records of the seventeenth century tell some interesting things about the sex ratio and about the age distribution between adults and children. K. G. Davies has analyzed 60,000 slaves delivered to the West Indies by the Royal African Company between 1673 and 1711; he reports that "51 per cent were men and 35 per cent were women; of the remainder 9 per cent were boys and 4 per cent girls. The dividing line between adults and children in slave tabulations is always vague. Generally, the sugar planters categorized black children up to about age ten as 'pickaninnies,' youths from about ten to fifteen as 'working boys' or 'working girls,' and anyone over sixteen as 'men or women.' " In any case, Davies's figures show that very few children were brought over from Africa. And though 86 percent of the arriving slaves appear to have been old enough for procreation, the 60 to 40 sex ratio in favor of men would hamper effective breeding. The low proportion of females imported into the islands might seem to explain why the traditional African family structures failed to replicate themselves in Barbados, but this explanation turns out to be too easy.[25]

Dunn's research demonstrates that the sex and age characteristics of the slave population began to change as soon as the blacks were led off the slave ships and put to work in the sugar plantations. The male preponderance among incoming slaves quickly melted away. Female slaves survived better in Barbados and righted the sexual imbalance. Even though the Royal African Company delivered 60 percent male cargoes, the women slightly outnumbered the men in the island's slave gangs.[26] On William Eastchurch's plantation in St. Thomas there were 19 males and 21 females.[27] Fishers Pond Plantation, one of the largest on the island, had 36 males and 46 females in 1668.[28]

Why did the male slave die faster than the female? Men may have suffered more from overwork, from malnutrition, and from fatal accidents in the sugar works. Males were far more likely than females to run away or rebel; they were far more frequently flogged, maimed, and executed.[29] While the slave plantation system bore heavily on men, it was even more unkind to infants and elderly slaves. A planter encumbered by numerous infant Negroes, nursing mothers, and decrepit old slaves could well find his plantation slowly being turned into an almshouse rather than the business enterprise he expected it to be. Some of these heads of households found it hard to resist the temptation to get rid of the young and old by systematic neglect and underfeeding.

If this judgment sounds too harsh, consider J. H. Bennett's description of Codrington Plantation. Six Negroes died to each one born at Codrington in the years 1712–48, a death rate more than six times as large as the birthrate. The typical year saw 2 Negroes born and 13 die in a plantation that had numbered 238 at the beginning of the twelvemonth period.[30] On the island in general, planters estimated that black infants had a 1 in 2 chance of living to the age of five.[31] The high mortality rates among slave children, and the fact that they could do little work on the plantation, caused planters to see infants and toddlers more as financial obligations than as assets. Indeed, in inventories of their plantations, planters rarely placed any value next to the name of a child until he reached the age of ten. The lease agreement between Timothy Thornhill and George Hirst is illustrative of planter indifference toward the young. In 1682, Thornhill decided to lease his Four-Hill Plantation in St. Peter to Hirst. On this 200-acre estate resided 91 slaves (32 men, 44 women, and 15 children). All the adult bondsmen were appraised in the lease agreement, but the 15 children, the document reads, "were not appraised being small and at present both unfit for work and also unfit to be taken from their parents." They were therefore to remain on the estate, and Hirst was to have whatever work he could get out of them for "their accommodations." For his part, Thornhill agreed to accept the loss if any of the children died before the lease expired.[32]

Planters were generally aware that their slaves were not reproducing their numbers. In 1689, Edward Littleton, who had 120 slaves, wrote that the master of 100 slaves had to buy 6 new ones a year in order to maintain his stock.[33] If this was true, a slave on Littleton's plantation had a life expectancy of seventeen years. Calculating from past experience, Henry Drax estimated that he could anticipate 10 to 15 deaths among his working blacks per year—a mortality rate in excess of 40 per thousand.[34] When writing down his expectations, Drax was referring only to his prime laborers, excluding the elderly, infants, and small children. Their inclusion would have pushed slave mortalities on his estate well above 60 per 1,000.

Drax was not a careless planter when it came to providing for the physical needs of the slaves. He well understood that a plantation whose workers were weak and sickly would soon encounter financial difficulties. He thought it was "far better, making slaves to go through their work with cheerfulness," that bondsmen be given "plantations provisions enough."[35] Yet, for all his concern, long working hours, inadequate diet, lack of interest in the elderly and infants, and the demands of a plantation economy translated into high death rates for his blacks.

With slaves dying faster than they could reproduce their numbers, and planters having to buy replacements, the complex African kinship network could not take hold in Barbados. When a child managed to survive infancy, he matured without the "mothers," "fathers," "aunts," and "uncles" that surrounded the young in Africa. For Barbadian blacks African notions of kinship and lineage lose much of their social importance.

Demographic patterns were not the sole reason why the African family unit was not reconstituted in the colony. The priorities and the overall structure of the plantation household ensured that the slave family was not a replication of an Old World institution. An incident that occurred between Macow and his master, Thomas Modyford, is a revealing illustration of how African familial values were undermined in Barbados. Apparently, Macow came from an African community that believed that the birth of twins was evidence of a wife's unfaithfulness to her husband. The punishment for such infidelity was hanging; thus, following ancient tradition, Macow was determined to hang his wife. Informed by one of the Christian servants of the slave's intention, Modyford tried to persuade Macow that twins were not a sign of unfaithfulness, but the argument had no effect upon him whom "custom had taken so deep an impression." Unable to change Macow's convictions, Modyford told the slave that if he killed his wife "he himself would be hanged by her." The threat had its desired consequence; Macow left her alone, and as a reward Modyford provided him with another mate.[36] The episode is reflective of the general drift away from African norms that occurred in the colony. Planters used their authority to decide what was proper social conduct in their households. In this new environment the African's sense of self was dramatically altered.

A majority of the Africans brought to Barbados were sold to one of the large sugar estates. On these plantations, they encountered other unfortunate souls from areas contiguous to their homeland. Where mutual intelligible languages and no tradition of hostile associations existed, "new Negroes" organized themselves into a particular ethnic constituency within the planter's household. They were their master's Gold Coast slaves as opposed to his Congo, Papaw, or Gambian Negroes. These Africans were basically assuming new ethnic identities. In Africa there was no sociopolitical entity that embraced all the peoples from the Gold Coast. In that region of West Africa, individuals spoke of themselves as being a Fanti, a Ga, a Guang, or one of the Akan-speaking peoples. Even in Western Central Africa, where the Mani-Kongo had managed to conquer the plateau of Longo, there was no all-embracing ethnic consciousness that bound the Teke, Lunda, Kuba, or Kongo peoples together. Unable to reconstitute their traditional communities in the New World, individuals from the same approximate geographical area in Africa developed a sense of commonalty in Barbados. Fanti, Ga, Guang, and Ashanti referred to themselves as Gold Coast Negroes, while Teke, Kuba, and Lunda adopted the name Congo blacks, meaning by that a person from that region in Africa.[37] Recent research on the eighteenth and nineteenth centuries suggests that where possible "new Negroes" tended to marry with those from areas contiguous to their homeland, and they arranged their thatch huts so that they formed ethnic enclaves on their master's estate.[38] By the close of the seventeenth century, as the rebellion of 1675 makes plain, the new ethnic identities Africans had come to assume were beginning to emerge as a political force in the colony.

The revolt involved only African-born male plantation slaves. "It was hatched by the Coromantines or Gold Coast Negroes," who, according to Governor Jonathan Atkins, "are much the greater number [in Barbados] from any one country, and are a warlike and robust people."[39] Englishmen in the colony regularly spoke as if Coromantine slaves were all from one community in Africa, though some of them undoubtedly knew that such was not the case, and Atkins was just reiterating local convention when he described Gold Coast Negroes as being from one country. The truth is that these slaves represented various groups of African peoples who were shipped to the island from Fort Coromantine. The consensus among Barbadian planters was that Africans transported from this fort made the best bondsmen.[40]

Encouraged by planters, Gold Coast Negroes throughout the island soon conceived of themselves as being part of a separate, elite group among "new Negroes," and in 1675 they put together a revolt that "spread over most of the plantations." They plotted for three years for the night when they would set fire to the canes and "cut their masters' throats in their respective plantations." The design was to "kill all whites in a fortnight." Once they had taken over the island, the Gold Coast Negroes planned to "choose them a king, one Coffee, an ancient Gold Coast Negro who should have been crowned on the 12th of June 1675 . . . in a Chair of State exquisitely wrought and carved after their mode; with bows and arrows to be likewise carried in state before His Majesty, their intended King."[41]

Coffee was probably from one of the Akan-speaking communities on the Gold Coast. He no doubt was a charismatic leader who proposed new religious rituals which would bring fortune and invulnerability to his followers and which would overcome the witchcraft of planters. In any case, the "Chair of State," or stool, was of fundamental significance to Akan people as a symbol of political authority, and group permanence and identity.[42] Using this traditional political symbol, Coffee organized the admixture of African tribesmen, who called themselves Gold Coast Negroes in Barbados, into an insurrection that may indeed have been aimed at overthrowing the plantation system and replacing it with an Asante-type kingdom, as Jerome Handler recently suggested.[43] The reality of plantation life brought changes in the ethnic consciousness of Africans. They redefined group membership, thus providing people from distinct ethnic communities in Africa with a common identity. A part of the African cosmology (that understanding of how the world functioned phenomenologically) informed relationships within these groups, and so it is possible to trace political symbols of authority or religious practices back to their African roots. Yet, demographic changes and the rhythm of a plantation economy ensured that sugar estates were not traditional African communities. Inside the plantation household, African slaves came face to face with a world market economy that altered their sense of self.

In contrast to "new Negroes," Creole slaves did not form an allegiance to one of the African regional constituencies that emerged in the colony. Creoles believed that they were a class apart from "new Negroes." On large plantations, these blacks tended to form an elite class, serving as the skilled and managerial bondsmen in their master's household. Some of them had two or three wives, and many of them received a greater allotment of food and clothing than the "new Negroes" in the plantation gangs. They were equally conspicuous in Bridgetown and in small planters' households. Some Creoles enjoyed their owner's trust to the extent that in times of emergency, when the trainband was called out, they were temporarily recruited into the militia.[44] Planters gave large responsibilities to these acculturated blacks; yet, by the 1660s—roughly twenty years after the mass importation of Africans into Barbados began—planters began to complain about an increasing number of creolized slaves in their households.

Slaveholders' uneasiness proceeded from an opinion that developed in the 1640s that the best security was afforded when the slave population was divided along cultural and linguistic lines. After describing how the militia was used to prevent slave insurrections, and how efforts were made to keep weapons out of the hands of slaves, Richard Ligon went on to explain that the absence of revolts on the island could also be attributed to the fact that slaves "were fetched from several parts of Africa, who speak several languages, and by that means, one of them understands not another."[45] In Creole slaves planters saw this latter safeguard disappearing. They had acquired a basic knowledge of English, which they used to communicate not only with their masters but with each other as well. Equally unsettling because of their occupations, Creole slaves tended to have greater geographical mobility than the newly imported Africans. As cartsmen they freely traveled the island's roadways; in Bridgetown they walked the streets on errands for their masters without any immediate supervision. Individually, they were always showing signs of restlessness with their status, straining for advantages that would lead to some meaningful change in their social and economic circumstances. Collectively, they were evolving toward a new group consciousness whose boundaries were fixed by notions of race and class. With these considerations in mind, William Willoughby wrote home in 1668 that "although the different tongues and animosities" among slaves had inhibited their rebelling in the past, he feared that the "creolian generation now growing up and increasing may hereafter 'manicipate' their masters."[46] The insurrections of 1692 gave credence to their fears.

Much of what we know about the revolt was extracted from blacks while they were being tortured, and the reliability of such testimony is open to question. Unlike the conspiracy of 1675, "this plot was formed by the Negroes that were born on the island." A Barbadian court-martial empaneled to investigate the uprising concluded that "most of the chief officers" or principal leaders, "as also the greatest villains amongst them . . . are generally overseers [i.e., black drivers], carpenters, bricklayers, wheelwrights, sawyers, blacksmiths, grooms, and such others that have more favour shown them by their masters, which adds abundantly to their crimes."[47]

The goals of Creoles in this rebellion stand out in marked contrast to the objectives of the Gold Coast Negroes in 1675. The aim of the 1692 insurrectionists was to "kill the governor and all the planters, and to destroy the government . . . and to set up a new governor and government of their own." More specifically, "they designed to have taken up the surnames and offices of their masters."[48] The society they sought to control was not to be far different from what existed in Barbados. They intended that no "imported Negro [i.e., New Negro] was to have been admitted to partake of the freedom they intended to gain, till he had been made free by them, who should have been their masters."[49] In effect, these Creoles hoped to take the place of the slaveowners as the governing force in Barbados, without altering the basic fabric of plantation society. Theirs was not a vision of a return to a traditional African way of existence.

In Barbados, African ideas of kinship and lineage lost much of their societal significance, as the traditional African family structure failed to make the journey to the New World. African ethnic identities (Ga, Kuba, Ashanti, Ijaw) also lost their relevance in the island's slave plantation system. In the colony, it was more important to be a Creole slave as opposed to a Gold Coast Negro, or a person transported from Fort Coromantine as opposed to someone from Western Central Africa. Although these divisions in the slave community had social and political meaning, as the two rebellions plainly reveal, they should be viewed primarily as stratification within a single caste. While being a Creole or a Gold Coast slave had implications for one's status in Barbadian society, by the close of the seventeenth century the racial categories of black and white emerged as the most significant lines of social differentiation in the colony.

The breakdown in the religious stratification of Barbadian society was in part occasioned by the efforts of slaves. By the latter half of the seventeenth century, a few Creole bondsmen were sufficiently bold enough to disregard their masters' orders and seek admission in the church, putting planters in the uncomfortable position of making Christian martyrs of their slaves. Morgan Godwin described the fate of one Creole black who requested and received "baptism upon a Sunday-morning at his parish church." When the black returned home, his master instantly took him to task, informing him "that that was no Sunday work for those of his complexion." The slaveowner went on to say "that he had other business for him, the neglect whereof should cost him an afternoon's baptism in blood . . . as in the morning he had received a baptism with water."[50]

The death blow was dealt to the Christian/non-Christian polarization of Barbadian society when the Church of England ruled that conversion did not emancipate slaves, rejecting the theological view that one Christian could not enslave another.[51] The Anglican church was very much for the Christianization of blacks and worked through colonial officials in the mother country to bring it about in Barbados. Under pressure from imperial authorities, in 1691 the General Assembly halfheartedly approved a motion by the governor that some convenient way be found to baptize blacks.[52] This vote can be taken as a sign that paganism no longer distinguished a slave from a freeman in the colony.

Even with assurances that conversion would not free their slaves, Barbadian planters remained opposed to the Christianization of blacks. Their continued resistance was rooted in a growing racialist conviction that Negroes constituted a nonhuman form of life. "Any minister who dared to baptize them," a pious woman opined, "might as well baptize a puppy."[53] Negroes, another declared, "were beasts, and had no more souls than beasts."[54] The idea that blacks were beasts without redeemable souls was part of a larger, but not clearly delineated, belief system that categorized them as a suborder of man—"ape-man" was Henry Whistler's term.[55]

As Europeans in Barbados became more deeply involved in slavery, they promoted rituals in their households to signify and reinforce the lowly status of bondsmen. The removal of the hat and a complaisant tone of voice or facial expression were part of an elaborate system of deferential behavior that blacks were required to exhibit in the presence of whites. Ignoring the underlying physical intimidation that secured the master-slave relationship, it did not take long before some assumed that the condition and conduct of these bondsmen were in some way related to their race. European children nurtured in this poisoned environment came to see themselves and other whites as superior to the enslaved Negro. Thomas Jefferson described it this way:

> The whole commerce between master and slave is a perpetual exercise of the most boisterous passions, the most unremitting despotism on the one part, and degrading submissions on the other. Our children see this, and learn to imitate it. . . . The parent storms, the child looks on, catches the lineaments of wrath, puts on the same airs in the circle of smaller slaves, gives a loose of

his worst passions, and thus nursed, educated, and daily exercised in tyranny, cannot but be stamped by it with odious peculiarities.[56]

One of the odious viewpoints the children of early Barbadian planters absorbed was a sense of superiority to people of color.

The rise of racial consciousness in Barbados was not simply a matter of planters and their offspring learning to perceive Negroes as an inferior race whom nature had given them a right to dominate. As has been mentioned, there was a widespread fear among whites, at times verging on hysteria and paranoia, that slaves might accomplish a successful revolt that would result in their massacre, and this contributed to the evolution of a racialist mentality on the island. Rather than link the social volatility of slaves to the contradictions that existed in their households, planters made it appear as if there was some inherent racial quality that made blacks socially unstable. One of the island's first slave codes describes blacks as a "brutish and uncertain dangerous" people who had to be governed by special laws.[57] Ligon accounted them a "bloody people, where they think they have power or advantage," so that he felt that planters always had to keep the upper hand over them.[58] It was these fears of a slave insurrection that helped to spur on ideas of racial solidarity among Europeans.

The European community in Barbados was far from homogeneous, as the presences in the colony of English, Dutch, French, Scots, and Irish would suggest. Yet, in its dealings with blacks, the idea of being white became a transcending, binding force that elicited racial solidarity. In effect, the social conditions that were a by-product of the slave plantation system bred a racialist mentality among European settlers that caused them to perceive blacks as a dangerous but inferior race who should be dominated by whites. By the second half of the seventeenth century, an ideology of white supremacy was budding in the island's popular culture. Some attempted to rebut racialist thought even in these early years of its development (most notably, Morgan Godwin and Thomas Tryon), but most planters found a pleasing ring of truth in it.[59] Although it would take a later generation to provide the "scientific," high-culture trappings, by the 1690s patriarchs were using racist contentions to explain and justify their behavior.[60] They argued that the subjugation of blacks in their household was in keeping with the natural order of the world. Governor Atkins put it most succinctly when he told the island's General Assembly that "God and nature" had universally designed blacks for slavery.[61] With men of Atkins's persuasion in power, it is easy to see why the government pursued policies that ensured the survival of the plantation household and why no one protested when an act of the General Assembly excluded all blacks from participation in the island's political system. The words "Negro" and "slave" had by "custom grown homogeneous and convertible" in Barbados.

While the assertion by planters that blacks were inferior to whites had the desired effect of rationalizing the systematic exploitation that was occurring in their households, by establishing slavery on somatic differences they also created a permanent underclass whose basic interest could never be accommodated in a slave plantation system. Planters understood this fact, and because of it they realized that in an emergency few of the island's blacks could be depended upon to defend the social order. The racial stratification of Barbadian society left the colony vulnerable and dependent on English support.

PART III

CONCLUSION

CHAPTER 10

Slavery as a Sociopolitical Problem

When modern historians write about plantation societies, they do so with great sensitivity to the moral questions that slavery posed. Less frequently do they raise the possibility that serious social and political problems were endemic to these communities. This tendency is perhaps explained by their reliance on pre–World War II political narratives. These earlier studies say very little about the plight of the average man and his place in the political economy of plantation societies. It is at least clear in the instance of Barbados's slave plantation system that inner tensions and strains had long-term consequences for the colony's political development.

The Barbadian political economy was deeply influenced by the widespread use of slave labor, as any seventeenth-century English political thinker could have anticipated. To be sure, as this study has sought to demonstrate, the drift away from traditional English political values and norms began well before the mass importation of slaves into the colony. It started in those first years of settlement when planters established their right to govern the dependent laborers in their household with little interference from church, state, or community. By the close of the 1630s a small plantocracy owned a third of the island's population, whom they kept on their plantations as servants and slaves.

Finding tobacco unprofitable, in the late 1630s some Barbadian plantation owners began experimenting with sugar. The success they had with the new staple brought a crazy excitement to the colony. Shipload after shipload of new settlers landed on the island anxious to have their turn at the wheel of fortune. Without giving much consideration to the consequences, planters imported African slaves to produce sugar on their estates.

In Barbados, Africans and Europeans overcame the social and linguistic barriers that stood between them and forged new relationships and institutions. On a daily basis much of the contact between the two groups occurred within the plantation household. The structure of the plantation household, with people of African descent in perpetual bondage and Europeans their masters, created enduring contradictions that threatened the existence of the institution. While planters effectively utilized the powers at their disposal to arrest the tendency toward instability in their households, ultimately the local government had to play a role in the perpetuation of the master-slave relationship.

As a rule, the state tried to interfere as little as possible in matters related to the governance of slaves. Whenever it became involved, its major concern was public safety. In general, the island's government viewed bondsmen as a population to be controlled rather than as citizens who might have to be called upon in an emergency. Although some trusted blacks were allowed to join their masters when the militia was answering an alarm, slaves were generally not thought fit to be recruits for the trainband. By 1680, 80 percent of the colony's population fell into the latter category.

The inequities that slavery produced were at first legitimated on religious grounds, but by the end of the seventeenth century, the theological justification for the enslavement of Africans had all but disappeared, replaced by notions of black inferiority. In their dealings with Negroes, Europeans saw themselves as a single ethnic group that nature had intended as governors for people of color. By this logic, blacks were consigned to be a permanent underclass that had no political rights. Planters recognized that the condition of slavery left bondsmen with frustrations that could have served as fuel for a servile insurrection.

The impact that racial slavery had on political relations in Barbados went well beyond the potentiality of a slave revolt and the need to reduce the likelihood of such an occurrence. Because the island's economy made almost exclusive use of slave labor, free workers found little opportunity for employment in Barbados; as a result, few free inhabitants can be identified as having a genuine material interest in the community. In the seventeenth century, the majority of the islanders (both black and white) were alienated from the society to the degree that big planters had to realistically question whether they could be depended upon in a time of crisis. By preying on the fears and superstitions of men, the plantocracy was able to use racial and religious differences, a selective dispensation of material rewards, and sheer physical intimidation to solve many of the problems related to internal tranquility. But to survive in the competitive and contentious environment of the seventeenth-century Caribbean, the plantocracy was also obliged to look to England for military protection against external aggression. England gave the islanders the assistance they required to maintain the integrity of their society—but at a price: the colonists were expected to defer to the will of the mother country.

For a brief moment, during the height of the sugar revolution, planters tasted local self-government. This autonomy was founded upon the breakdown of the English central government in the 1640s. In 1649, Parliament, victorious over the king, brought some semblance of stable central government back to the mother country. Soon it was pressing Barbadians to obey its directives. Planters did all they could to preserve home rule—even to the point of taking up arms. In the struggle, plantation owners learned the bitter lesson that the social organization of their society could not support their desire for independence. They surrendered to England.

After the capitulation, the colonists tried to salvage some portion of their former autonomy by arguing that the island should be incorporated as an English town or borough with the right to select its own local officials and the right to elect representatives to Parliament. Had the House of Commons agreed to this proposal, the history of Barbados, indeed of Anglo-American relations, could have taken a very different course. At the time, however, a mercantilistic view of the empire held sway in England, and the English elite were quite prepared to regulate their New World possessions as dependent colonies.

Throughout the second half of the seventeenth century the metropolis and the colony habitually quarreled about some facet of the colonial system—rights of the Assembly, free trade, the appointment of public officials were dominant issues. In those areas where the interests of England and Barbados collided, planters' needs generally took second place to those of the mother country. The rise of patent officers, the colonial acceptance of the Navigation Acts (by which England regulated the island's trade), and the general stability of the imperial structure must be understood in relation to the social and political problems that slavery created for the colony's governing elite.

If members of the plantocracy had built a social order more compatible with their home rule ambitions, there would have been no guarantee that the island would have been able to survive as a free state. Its small geographical size and its reliance upon shipping to link it with its European markets fostered additional weakness that could have also frustrated their desire for local autonomy, but whether these latter factors raised insurmountable obstacles to independence must be left to the realm of counterfactual speculation. What is certain is that one of the basic prerequisites for home rule was absent in the island's slave plantation society—a local population who would have willingly sacrificed to maintain the island's independence. Planters recognized this fact, and there is no evidence of them reasoning that since the island was too small to support a free state then they might as well extract the largest possible return from their bondsmen.

When compared with other plantation societies in the New World, the internal problems and instability that the Barbadian slave economy produced do not seem atypical. In Virginia, unlike Barbados, writes Edmund Morgan, poor whites did not migrate out of the colony but drifted into the hinterlands. Faced with an ever expanding poor white population, and threatened by the possibility of a slave revolt, Virginia slaveowners courted the favor of all whites in common contempt for persons of a dark complexion. Believing their own propaganda, the ruling elite, having proclaimed that all white men were superior to blacks, offered their social (but white) inferiors a number of benefits that improved their lot in the society. Thus, instead of the avarice of their superiors squeezing them out, as in Barbados, Morgan writes, men of small means were allowed not only to prosper "but also to acquire social, psychological, and political advantages that turned the thrust of the exploitation away from them and aligned them with the exploiters." In comparison to Barbados, this alliance provided the colony with a relatively large number of inhabitants who would support the social order.[1] The sizable free white population in the American South did not bring total immunity to the vulnerabilities that racial slavery engendered, as was evident during the American Revolution.

The basic demographic facts in 1776 were as follows. Ninety percent of North America's blacks lived in the five southernmost colonies (Virginia, South and North Carolina, Maryland, and Georgia), where 95 percent of them were held in slavery. Forty percent of Virginia's population was black. In South Carolina, Negroes and whites were roughly equal in number, but in the districts nearest the sea, those most susceptible to attacks from the ocean, blacks outnumbered whites by as much as 10 or 12 to 1. No one knew how slaves would react to the coming of the war to their neighborhood, but everyone appreciated that the outcome might depend on the answer.[2]

During the first winter of the war, the British expected that loyal colonial governments could be restored in Georgia and the Carolinas by a single vigorous assault. Their strategy was based on the assumption that most southerners were still loyal to the king and that the presence of a large number of slaves left these colonies so weak that a strong show of force would be enough to embolden Loyalists to seize power and exercise it in obedience to the British crown. Governor Josiah Martin of North Carolina contributed to this sense when he reported that Britain had many friends in the colony. He further added that the British need not fear the leaders of the rebellion in North Carolina, most of whom lived on plantations in the Tidewater region; they were outnumbered by their slaves, and were thus effectively neutralized.[3] Citing Martin's report, as well as others from Virginia and South Carolina, to the same effect, Lord North, the British wartime prime minister, commented that "the accounts we have received from these provinces are the more to be credited as we all know the perilous situation of three of them from the great number of their Negro slaves, and the small proportion of white inhabitants."[4]

The patriots knew that if the British decided to mobilize slaves their cause could be placed in serious jeopardy. A delegate to the Continental Congress from Georgia told John Adams in 1775 that the British could subdue Georgia and South Carolina in a fortnight, with a mere 1,000 regular troops, simply by promising freedom to the Negroes there. Twenty thousand Negroes would flock to the imperial standard in response to such a declaration. The only force that prevented the British from taking such a course, he surmised, was that Loyalists also had Negro slaves, and would surely disapprove of such a policy, since it would be impossible to confine the disruption to the rebels.[5]

Slaves were the great unknown quantity in the military balance. Men on both sides owned large numbers of them. If either side had been prepared to pronounce a general emancipation, it might immediately have had a large number of black men available for military duty. But each side was restrained from pursuing such a course by the social revolution that would necessarily have accompanied it.

In November 1775, Lord Dunmore, governor of Virginia, began prying at Pandora's box when he promised freedom to slaves who would repair to the king's standard and "bear arms" against the rebels. Dunmore confined his offer of emancipation to slaves "appertaining to rebels," but the alarm among slaveowners in Virginia and throughout the colonies was general.[6] George Washington had this to say of Dunmore's action:

> if the Virginians are wise, that arch-traitor to the rights of humanity, Lord Dunmore, should be instantly crushed, if it takes the force of the whole army to do it; otherwise, like a snowball in rolling, his army will get size, some through fear, some through promises, and some through inclinations, joining his standard: but which renders the measure indispensably necessary is the Negroes; for if he gets formidable, numbers of them will be tempted to join who will be afraid to do it otherwise.[7]

Dunmore's proclamation could have been a more decisive event in the war, turning the balance of power to the imperial side, but to capitalize on the southern weakness certain changes in the attitudes of British policymakers were required. First, they had to overcome their squeamishness about the place of blacks in the British army. Second, they had to bring themselves to the point where they could convince skeptical slaves that they would be safe in British arms and that they would not be sent off to a harsher slavery in the West Indies once their military service was done. Most important, the British had to decide what they intended for the future of America and how Negro slavery fit into that future. If Britain were to fully capitalize on the opportunity presented by the bondage of blacks, it would have had to accept a revolution in the southern social system—something that in the last quarter of the eighteenth century it was not prepared to do.[8]

The British had failed to press their advantage, but a more ruthless and resourceful foe might have inflicted greater punishment on the South—a point that was not lost on the victorious patriots. During the Constitutional Convention of 1787, a debate broke out about the wisdom of allowing slavery in the new Republic. George Mason reminded the convention that the evil of having slaves was experienced during the War of Independence. Had the slaves been treated as they might have been by the British, they would have proved dangerous instruments in their hands. Using history to support his case, he recalled the dangerous insurrections of slaves in ancient Greece and Sicily, and the instructions given by Cromwell to the commissioners sent to reduce Virginia in 1652—to arm the servants and slaves, in case other means of obtaining its submission should fail.[9] Had he known of Barbados's experience in the 1650s, he could have added that colony to his list of historical examples of the evil political consequences of slavery.

Looking at the Western Hemisphere in general, there is good reason to believe that planters in Dutch, French, Hispanic, and Portuguese colonies were also confronted with social, political, and military difficulties that bore a resemblance to those faced by Barbadian and North American slaveowners. In his assault on Panama in 1572, Sir Francis Drake found many Hispanic slaves quite willing to help him attack their masters, looking upon him as a liberator.[10] The comparative slavery debates reveal that in many of the plantation communities in Latin America and the Caribbean there was a chronic shortage of European freemen who might be inducted into the militia. The privileges of free blacks in these societies were often predicated on their usefulness as substitutes for whites in the militia.[11] The Haitian Revolution stands as indisputable evidence of how a contest between colonials and imperial authority in a slave plantation society could quickly involve slaves with revolutionary consequences.[12] Unlike North America, in many Latin American countries the emergence of independent nation-states was often closely related to the death of racial slavery. Simón Bolívar in his campaign to topple the Spanish colonial system frequently recruited slaves by promising them freedom, a fact that was directly responsible for the end of servitude in Venezuela and Colombia.[13] In the Cuban struggle for home rule, the rebels also offered emancipation to bondsmen in exchange for their assistance against the imperial power, and so they helped to abolish slavery on that island.[14]

Slavery left an indelible mark on the political economies of plantation America, and we need more comparative studies before we can fully appreciate the depth of its impact. Perhaps of more immediate concern to this generation, it also left a legacy of inequities and racial myths that we can hope to transcend only by acquiring an informed sense of their origins.

APPENDIX

White Demographic Patterns in Colonial Barbados

Insofar as the history of plantation communities in the Caribbean is concerned, it is generally recognized that the plantation system had an effect on population movement. It is commonly accepted, for example, that the ability of slaves to reproduce their numbers was in part determined by the dynamics of a plantation economy. Most historians would also agree that the structure and composition of the white population was also shaped by the internal workings of these communities. In his pioneering study of the British West Indies, Frank Pitman outlines how the conversion to slave labor and the monopolization of land by a few planters in the mid-seventeenth century triggered a pattern of migration away from Barbados that inhibited the growth of the white population. Pitman supports his thesis mainly with eighteenth-century sources, making no serious efforts to study the preceding century to test the validity of his claims.[1]

Richard Dunn, in his *Sugar and Slaves*, is the first to really challenge Pitman's interpretation. Through his work on surviving census material, a source rarely used by other historians, Dunn breaks from Pitman and concludes that the inability of whites to increase in the West Indies was the product of a demographic disaster caused by high mortality. The Englishmen who settled in the Caribbean, Dunn calculates, "shortened their life expectancy significantly." The mortality rate was so high that although they "married in the islands and raised families, they did not increase and multiply." In order to sustain their numbers, they "depended on constant in-migration." Early death, Dunn believes, retarded the family as an institution of socialization. Compared with New England where parents "trained their children in habits of restraint and introspection," the West Indies colonists "lacked effective family guidance and were more undisciplined, exhibitionist, and freewheeling." In Barbados, the parishes were thrown into continuous flux as death, or the fear of it, bore down upon the inhabitants, making it impossible to produce the New England "stable, cohesive little communities, where children habitually settled near their parents and few planters moved away." Moreover, early death ensured that there were no elders on whom political stability might be based, leaving the young in control. The specter of an untimely end explained the "frantic temper and mirage-like quality of West Indian life—gorgeously opulent today, gone tomorrow." Whites longed for an "early retirement in England," Dunn maintains, because "it was impossible to think of the sugar islands as home when they were such a demographic disaster area."[2]

In Pitman and Dunn we get two very different views of how Europeans fared in seventeenth-century Barbados. One has them migrating away from an economy in which few could prosper; the other sees a physical environment too infested with pathogens for the settlers to survive in. It would seem that another look at the question would be appropriate.

For the presugar years there are no census data, vital statistics, or militia rolls that might form the basis of a quantitative analysis of population development in Barbados. In the absence of numbers, we rely on literary sources. This material makes it plain that it was the rare occasion when the incidence of death in the colony reached a proportion sufficiently unusual as to provoke a comment from the colonists.

In the early years of settlement in Virginia, death was omnipresent, and the colonists frequently spoke of its afflicting hand. Richard Frethorne was merely echoing the opinion of many pioneers when he wrote home that the nature of Virginia was "such as caused much sickness;

as the scurvy and the blood flus, and divers other disease, which make the body poor and weak."[3] The high incidence of death in South Carolina caused similar lamentations.[4]

In the letters, journals, diaries, and other records of seventeenth-century Barbados, one commonly encounters such passages as "our bodies having been used to a colder climate, find a debility, and a general failing in the vigor and sprightfulness we have in colder climates, and our blood too, is thinner and paler than in our own countries."[5] For men who perceived the world through humoralistic eyes, the Barbadian climate and the strangeness of their diet were in themselves pathogenic. Yet, for all their misgivings about the physical environment, the fact remains that Barbadians were not living with morbidity and mortality rates approaching those of early Virginia or South Carolina.

The healthiness of Barbados is no doubt related to the absence of lakes and rivers on the island. The general shortage of water meant that many of the bacteria, protozoa, and viruses that today we associate with "tropical diseases" found the Barbadian ecology inhospitable. Malaria, for example, is caused by a protozoan of the genus *Plasmodium*. These microscopic parasites spend part of their lives in the red blood cells of human beings, but they also must spend part of their lives in female anopheles mosquitoes, which carry and spread the malaria infection. The absence of lakes and swamps in seventeenth-century Barbados prevented the growth of an anopheles mosquito population, and without the insect there was no possibility of a malaria outbreak in the colony. The first malaria epidemic in Barbados did not occur until 1924.[6]

The *Aedes aegypti* mosquitoes, which carry the yellow fever virus, would have found the barrels and cisterns that planters used to collect rainwater an adequate breeding place, but the frequent periods of drought meant that these receptacles could provide them with only temporary homes. This fact would seem to explain both why yellow fever epidemics occurred in the colony and why the disease never became endemic to the island.[7]

In 1647 the colony experienced its first health crisis. Apparently it was part of a yellow fever epidemic that was sweeping the West Indies in the late 1640s. It would seem that the *Aedes* mosquito was being spread in the Caribbean by ships trading with the various colonies. The mosquito certainly would have been attracted to the barrels that seamen used to store water. Through these means the mosquito made its way to Barbados, bringing with it the yellow fever virus. Richard Vines, who had only recently settled in the colony from New England, reported that in one parish as many as 20 people had died in a week, and in many weeks 15 or 16 parishioners had succumbed to the disease.[8] Governor John Winthrop of Massachusetts estimated that a total of 6,000 Barbadians had been killed by the pestilence.[9] Faced with the prospect that some of his people might move to the West Indies, the governor had some reason to exaggerate the extent of the illness. Yet, the epidemic was serious. Richard Ligon, who arrived in the colony in time to witness the ravages of the disease, wrote that within a month "the living were barely able to bury the dead."[10] Fortunately for the colonists, the outbreak did not last long. In Barbados, the *Aedes* mosquito was introduced into the colony just after a serious drought, and the shortage of water probably prevented the insect from reproducing itself. By 1648 reports of the epidemic ceased, and there were no further accounts of unhealthy times in Barbados until 1671.

The general wholesomeness of Barbados in these early decades makes sense when one considers that the white population grew from 6,000 in 1638 to 18,000 by the mid-1650s, helping to make it the most populated colony in the British Americas.[11] After the sugar revolution, planters were inclined to use slave labor in their households, and they were generally indifferent to the plight of free workers who found it difficult to support themselves in the island's plantation economy.

In their reactions to the many natural disasters that befell the island in the seventeenth century, the Barbadian government repeatedly showed a reluctance to make any economic sacrifices that might have improved the lot of free laborers and encouraged them to stay on the island. The year 1640 was a "very unseasonable year": the island suffered from drought conditions. The inhabitants were therefore unable to harvest enough cotton to "buy necessaries for their servants," and with all provisions in short supply "most of the people were in danger of being famished."[12] In the summer of 1647, "there was so great a drought, as their potatoes and corn, etc., were burnt up."[13] To relieve themselves, Barbadians purchased supplies from New England. While conservation might help during a drought of modest length, it could do nothing to alleviate the effects of flooding. In 1670 heavy rains lasted seven months, followed by a crippling three months of "very dry weather." The ground began to "gape" as if it would devour the islanders. This, along with a "plague of caterpillars," produced a great dearth "even to famine of corn and potatoes." The common view was that the people were "like to endure cruel famine for several months."[14] As if the extremities of rain were not sufficient, there were also hurricanes to fear. In the seventeenth century the colony was visited by no less than five severe tropical storms, which brought devastation and turmoil in their wake (August 19, 1666; October 7, 1670; August 31, 1675; August 13, 1694; and October 7, 1694).[15] Reporting on the visitation of 1675, Governor Jonathan Atkins exclaimed that he had never "seen such prodigious ruin in three hours; there are three churches, a thousand houses, and most of the mills of the Leeward thrown down; two hundred people killed, and whole families buried in the ruins of their houses." He estimated that £200 would not repair the damage caused by the storm.[16]

These natural disasters impacted on the size of the island's white population. In the aftermath of a meteorological upheaval, with crops and livestock destroyed, or injuries to their servants and slaves, small and middling planters must have been pushed close to insolvency. Where were they to find the cash reserves to get them through the droughts of 1640, 1647, 1651, 1663, 1667 and the excessive rains of 1656, 1669, 1670 (not to mention the five hurricanes)? The hazards were magnified if they committed themselves to the cultivation of sugar, which required from fourteen to eighteen months' growing time, during which period the cane was completely at the mercy of nature. Even if they were less adventurous and planted a staple that matured earlier, a succession of times of unseasonable weather might still have spelled their undoing. The 1675 hurricane left planters "in such consternation and distraction that they resolved never to build again but leave the island." After reflection, some changed their minds and began rebuilding and repairing as rapidly as possible. There were, however, a "great many" who could not afford to rebuild.[17] Presumably they sold what was left and turned to new horizons where fate might prove less unkind; at least, that was the opinion of Governor Lord

Vaughan of Jamaica, who expected "great numbers to leave, and some of the best quality from Barbados."[18] Sir Joseph Williamson, secretary of the committee of trade and plantation, concurred with the governor's judgment. In a letter to Lord Vaughan he expressed his apprehension that, "from the accounts of the late ravages at Barbados by hurricanes," Vaughan "must provide ere long for good numbers of families from the island."[19] Barbados might not have experienced a loss of population through migration in the wake of natural disasters had the government tried to assist small and middling planters to recover from the damage. But it remained inactive, leaving many with little choice but departure from the colony.

Richard Dunn stresses a high mortality rate for the failure of Englishmen to increase their numbers in the colony. His case rests on the censuses of 1680 and 1715. Attempting to account for the gradual reduction in the number of whites in the closing stages of the seventeenth century, he points to the census of 1680 when in eighteen months "506 more whites died than were born." In the parish of St. Michael, "107 infants were baptized," while during the same period "82 infants and children were buried." In this parish "four persons were buried for every one baptized." Only three of the remaining ten parishes recorded a "surplus of births over deaths." From this Dunn concludes that "Englishmen in Barbados were not maintaining the existing level of population," implying that mortality was responsible for this trend.[20]

Dunn found corroboration in Patricia Molen's purely speculative remarks on the 1715 census and in his own inspection of the returns of Christ Church parish.[21] The Barbadian population in 1715, Dunn rightly observes, was "exceedingly youthful." The median age for both sexes was nineteen; only 16 percent of the inhabitants "had passed their fortieth birthdays, and only three percent had passed their sixtieth birthdays." The population, however, did not grow "in the way most young populations do." "Given the high mortality rate," Dunn writes, marriages were short. Both censuses (1680 and 1715) were pockmarked "with widows, widowers, and orphans." As a consequence, family life remained stunted.[22]

In presenting the finding from the 1680 census, Dunn could have described his analysis a bit more. We are not informed, for instance, of how he derived the figure of 107 infants baptized in St. Michael.

Tucked away in a footnote we are told that "not all baptized persons were infants; a few were older children and adults."[23] This remark does not make the matter clearer. What the records specifically say is that "William ye son of William and Margaret Stickland," or "Elizabeth ye daughter of John and Elizabeth Gosnel" were baptized.[24] No age is given. Needless to say, "son" and "daughter" do not imply age; nor can infancy be inferred from them. If we closely followed his example for the parish of Christ Church, we could be enumerating "Job the son of Walter and Dorothy Hart of about 22 years of age," or "Charity the daughter of Walter and Dorothy Hart of about 19 years of age," as well as "Lucy of the age of 2 years and six months, and Anthony ten days old, daughter and son of Anthony and Mabell Harris."[25] This may seem like peevish quibbling over minor details, but the point becomes critical in light of Dunn's statement that "82 infants and children were buried in St. Michael." In the case of baptism there is legitimate reason to suspect that the majority of sons and daughters were infants and children—except for Anabaptists, who would have gone unrecorded in these registers anyway—that was generally the age when this rite was performed. But death cares not about age. Thus, there is absolutely no reason to assume that deceased sons and daughters listed without ages were "infants and children."

Even if we accept these numbers, and include with them Dunn's average of 542 births and 562 deaths that are reported for the ninety-three years between 1710 and 1803, which suggests to him that the islanders "were learning how to survive better in the tropics," the efforts to document the existence of a demographic disaster seems meager.[26] At no time are we offered a crude mortality rate so that we might better appreciate the frequency of death in Barbados. Simply to tell us that 506 more people died than were born, or that 107 infants were baptized while 82 infants and children were buried, is not documentation of a staggering death rate that hindered population growth. These bare numbers are uninformative. It is quite possible that the determining factor was an unusually low birthrate, or that migration was high coupled with an average death rate, comparatively speaking. Indeed, had the vital statistics been arranged in a more comprehensive fashion, Dunn might have been less willing to place so heavy an emphasis on death.

Appendix Table 1 presents a crude burial rate derived from Anglican burial registers. Although the omission of sectarians from these registers produces rates that are lower than the actual burial rates for all white inhabitants of Barbados, dissenters constituted a tiny fraction of the colony's population. Thus, their absence should not radically affect the rates in Appendix Table 1.[27] A more difficult matter to judge is the number of people who were not included in these rolls for reasons less tangible than religious preferences. The discovery of new information on these unrecorded colonists might raise our crude burial rates somewhat, but it is hard to believe that the numbers would be substantially increased. Besides the unanswerable questions posed by the burial returns, the parish population totals that are used here to calculate the rates in all probability were estimates. Many of the lists were drawn up by churchwardens for each parish upon a request made to them by the island's governor. These were generally not official requests, so that the churchwardens were not obliged to make a correct count. The accuracy of their responses (and by that the rates in Appendix Table 1) is therefore contingent upon the personal diligence of the wardens.

It was not until 1678 that adequate vital statistics exist on which to construct a crude burial rate. There were four other occasions between 1678 and 1749 when the available information can be converted into a burial rate. It is possible to calculate burial rates for each year from 1749 to 1762. Appendix Table 1 is biased toward the mid-eighteenth century, when fourteen observations can be made compared with the five in the earlier years. Despite all its inherent problems and limitations, Appendix Table 1 is a refinement of the figures that Dunn presents as illustrative of consuming death. What the rates actually show is that Dunn's comparison of aggregated burial and baptism returns is not persuasive evidence of a high mortality rate.

Between 1678 and 1679 the average annual rate of burial was 30 per 1,000, and the median was 22. Of critical importance, these returns do not establish the presence of a demographic disaster of so great a magnitude as to determine the size of the colony's white population. Burials in 1683 seem to have reached an alarming level. In the absence of dates, the registers give no indication of how long a period they cover; the high rates may be a function of time. Unfortunately, there

APPENDIX TABLE 1

Crude Burial Rates by Parish (per 1,000)

Parish	*1678*	*1683*	*1708*	*1712*	*1716*	*1749*	*1750*	*1751*	*1752*	*Years* *1753*	*1754*	*1755*	*1756*	*1757*	*1758*	*1759*	*1760*	*1761*	*1762*	*Parish average rate for the periods surveyed*[a]
St. Michael	70	129	39	48	66	30	31	57	47	58	46	45	53	63	51	47	92	36	36	55
St. John	47	39	27	15	13	5	8	19	13	22	40	20	33	32	19	43	54	20	18	24
St. Philip	22	35	12	14	19	17	20	30	24	36	24	26	27	36	28	31	31	27	27	26
St. Joseph	21	48	*	11	13	17	14	14	26	28	25	11	15	17	23	16	65	21	15	22
St. Thomas	23	56	4	16	*	23	27	33	22	40	25	14	19	13	15	19	23	12	23	23
St. Lucy	20	46	*	22	27	20	22	39	18	30	40	28	18	48	37	26	40	16	18	29
St. Peter	*	57	24	17	41	21	26	28	21	34	25	19	14	26	34	22	38	13	26	27
St. James	17	50	37	29	27	13	22	17	23	39	30	30	35	27	24	29	27	19	25	27
St. George	31	51	32	31	28	30	51	45	38	88	65	19	55	67	42	46	77	30	52	46
St. Andrew	*	34	10	13	22	33	26	29	9	24	42	25	20	37	28	33	50	28	10	26
Christ Church	19	24	*	*	26	40	42	42	21	65	58	35	56	70	65	53	61	50	52	46
Barbadian annual average	30	53	23	22	28	23	26	32	23	42	38	25	34	40	33	33	51	25	27	

Sources: Sloane Ms. 2441 (British Library) F. 18; P.R.O., C.O. 1/44/47; C.O. 28/11/85; C.O. 28/14/2.I, 2.V; C.O. 28/15/16, 24.II; C.O. 28/29/Ee 9, Cc 39, Cc 135; C.O. 28/30/Dd 10, Dd 51; C.O. 28/30/Dd 62, Dd 75, Dd 100; C.O. 28/31/Ee 8, Ee 9, Ee 27, Ee 36; C.O. 28/32/Ff 2, Ff 15, Ff 26.

*No returns.

[a] Barbadian average rate for the period surveyed = 32.

are no burial statistics for the period between 1690 and 1709. From literary sources it is apparent that in these years the island experienced the worst health conditions in its history.[28] By 1709, however, the colony had recovered. The crude burial rates for 1715, based upon a census of that year which Dunn makes use of in his argument, do not support the thesis that family life was harmed by high mortality. The mean number of deaths per 1,000 was only 28.

The sources afford us a more detailed look at burials on the island between 1749 and 1762. One of the striking features of these rates is the radical fluctuation between parishes. St. John, for example, had the incredible lows of 5 and 8 per 1,000 in 1749 and 1750, respectively, and neighboring Christ Church whites were being buried at the rate of 40 and 42 per 1,000, respectively. In eight of the fourteen years studied, the Barbadian burial rate climbed above 30, with highs of 42 and 51. Although these data might seem to support Dunn's thesis, it should be noted that this increase is attributable to a smallpox epidemic that raged in the colony in 1752, and again between 1757 and 1760.[29] But even with the presence of smallpox, some parishes (namely St. Peter, St. Thomas, St. Joseph, and St. James) remained relatively healthy.

In spite of the sporadic upturn from 1753 to 1760, the mean numbers of deaths per 1,000 for the whole period surveyed (1678–1760) was only 32. Even this figure is deceptive. In three parishes—St. Michael, St. George, and Christ Church—the burials per 1,000 were well above those reported in the other eight. Their cumulative effect was to inflate the island's mean. If removed, the colony's average drops to the remarkable 26 per 1,000, and this includes the suspect rates of 1683. These rates suggest that Anglo-Barbadians were faring much better than colonials in the North American cities of Boston and Philadelphia, where the number of deaths in proportion to the total population in the eighteenth century was 37 and 42, respectively.[30]

From the construction of crude burial rates there are only three parishes that emerge as possible candidates for Dunn's demographic disaster area, though in none of these parishes did planters experience anything close to the 150 per 1,000 mortality rates that were commonly reported for Englishmen on the African coast.[31] In the case of Christ Church and St. George, it was not until the mid-eighteenth century that the burials began to approach a rate that could have in itself limited population growth. St. Michael was the only one that consistently returned a high burial rate. In the nineteen samples taken it had an average of 54. By comparing aggregate baptism and burial returns in Appendix Table 2, we get a better sense of how St. Michael might skew our understanding of demographic trends in Barbados. By removing the parish from the returns, baptisms drop from 11,508 to 8,731, a fall of 22 percent; burials decline from 11,376 to 6,685, a staggering 41 percent reduction. Failing to take stock of the

APPENDIX TABLE 2

Baptisms and Burials in Barbados[a]

Year	*Baptism*	*Burial*
1679	630 (523)	1,058 (637)
1683	407 (308)	1,026 (683)
1710	201 (183)	140 (100)
1712	449 (383)	352 (208)
1716	518 (412)	556 (254)
1723–26	1,543 (1,178)	1,434 (827)
1739	577 (431)	437 (217)
1748	366 (295)	483 (357)
1749	514 (389)	344 (210)
1750	569 (471)	399 (262)
1751	567 (420)	564 (313)
1752	396 (286)	442 (236)
1753	556 (433)	680 (424)
1754	543 (420)	638 (414)
1755	575 (437)	507 (290)
1756	588 (413)	584 (329)
1757	559 (428)	726 (423)
1758	583 (392)	611 (366)
1759	660 (477)	589 (361)
1760	682 (474)	960 (516)
1761	538 (393)	437 (261)
1762	524 (416)	493 (317)
Total	11,508 (8,731)	11,376 (6,685)

Sources: P.R.O., C.O. 1/44/47; Pitman, *British West Indies*, p. 385.

[a]The figures in parentheses are the same as those without except that the returns of St. Michael are deducted.

distorting influence that St. Michael had on baptism and burials statistics, Dunn perhaps underestimates how successfully settlers had adapted to their tropical environment. Even in St. Michael, where all figures seem to point to a high incidence of death, hasty conclusions should be avoided.

St. Michael embraces Bridgetown, the largest urban center in the colony and, more important, the chief port of call. Burials in the parish in all likelihood included, in addition to residents, new arrivals to the colony who might have contracted their fatal illness aboard unhygienic ships. The sources prohibit efforts to separate out new arrivals or transient mariners and passengers, but figures from 1739 are suggestive. In that year it was recorded that, of the 220 persons buried, 93 were "foreigners." What precisely was meant by that term is unclear, but it is likely that they were nonresidents of Barbados. The number of resident burials would then be the difference between the two, or 127. In 1739 "foreigners" accounted for 36 percent of the burials; residents, for 64 percent. There is no way to be sure if these percentages are representative of a general pattern, but they do indicate that nonresident burials could mislead historians. The matter is thrown into clearer relief if we calculate crude burial rates for 1739 using the unadjusted figure of 220 and then the 127 resident burials. The unadjusted rate per 1,000 was 50; adjusted, it fell to 29, a decline of 42 percent, which brought St. Michael burial rates well in line with those of other parishes on the island. If nonresidents consistently affected the burial statistics of St. Michael in anywhere near that proportion, it would certainly bring into question ideas that local planters found the parish unhealthy.[32]

Because the sources permit us to formulate only crude mortality rates for the closing decades of the seventeenth century and for the eighteenth century, what Appendix Table 1 hides is the fact that epidemics were less frequent in the first fifty years of settlement than in these later decades. In 1671 Wait Winthrop told his brother Fitzjohn that there had been a "great mortality in Barbados so that the people leave the island and go down to Nevis. They think it is something like the plague."[33] In 1680 Governor Atkins informed the Lords of Trade and Plantations that "it hath pleased Providence to send a great mortality among us these two years past which swept away many of our people and slaves."[34] Lieutenant Governor Stede wrote the Lords of Trade and Plantation in 1685 that Barbados was again experiencing a "very sickly time of fevers and smallpox, of which great number have died." Most of the illness in the latter epidemic was confined to Bridgetown, where the General Assembly refused to meet out of fear of sickness.[35]

The worst outbreak of illness occurred between 1691 and 1708. Commenting on this period, John Oldmixon, the early eighteenth-century historian, remarked that Barbados was once "reckon'd the healthiest island in America." This high opinion of the colony continued until 1691. In that year English troops carried to the island a "pestilential fever."[36] St. Michael was the worst hit. In 1694 there were reports of a plague in Bridgetown.[37] The summer of 1700 brought another epidemic to the island's chief port of call.[38] By no means, however, was all the pestilence located in Bridgetown. In 1703, Governor Granville notified authorities in England that the "island is more unhealthy than it was ever yet known to be." There was a "very dangerous distemper all over the country as well as in the towns." He described the illness as "very catching and very mortal."[39]

Regrettably, there are no burial returns between the years 1683 and 1716 that would enable us to measure the extent of the health emergency that occurred on the island. Oldmixon claims that in the thirteen-year period 1690–1703 fever killed almost a third of the colony's population, but this seems hardly credible inasmuch as the total number of whites was 17,230 in 1690, and it had dropped to 15,000 only in 1700. (Several historians have been deceived by the population returns of 1712 into thinking that only 12,528 whites inhabited the colony in that year. The 1712 totals, however, do not include the parish of Christ Church. If the parish had been counted, the returns of 1700 and 1712 would have been nearly identical.)[40]

By 1708 Barbados had regained its former reputation for healthiness. Governor Crowe told the English Board of Trade that the "island is generally more healthy than for many years before."[41] Oldmixon optimistically predicted that "the distemper is lately abated, and the colony increases daily, in which the present health of the place will, if it lasts, advance it in two or three years to the happy state it was in formerly."[42]

Throughout the remainder of the eighteenth century, Barbados remained generally healthy, though it never returned to the wholesomeness of former times, as Oldmixon speculated it would. There were minor epidemics in 1724, 1733, and the 1750s, but there was nothing to match the sickliness of the 1690s. Several new diseases, such as elephantiasis, appeared in Barbados in the eighteenth century.[43] The outbreak of these diseases provoked a considerable scientific crisis, as colonial physicians tried to employ humoralistic thought to explain their sudden manifestation.[44] Even with the introduction of new illnesses, the burial returns show that Barbadian whites never suffered under mortality rates similar to those that decimated Englishmen in early Virginia and South Carolina or on the African coast.

Eighteenth-century Anglo-Barbadians lived in an environment that was healthier than that of Boston or of Philadelphia in the same period. Yet, their numbers never increased beyond 20,000. Planters made a conscious decision (informed by classical medicine and economic needs) to employ black bondsmen as their chief source of labor on the island, and so there was little room in the colony's plantation economy to support a large free population. This fact more than anything else discouraged the growth of the white community.

Notes

PREFACE

1. Carl and Roberta Bridenbaugh, *No Peace Beyond the Line* (New York: Oxford University Press, 1972); Richard Dunn, *Sugar and Slaves* (Chapel Hill: University of North Carolina Press, Institute of Early American History and Culture, 1972); Jerome S. Handler, *Plantation Slavery in Barbados* (Cambridge: Harvard University Press, 1978), and *Freedmen in the Slave Society of Barbados* (Baltimore: The Johns Hopkins University Press, 1974).

2. Nicholas Darnell Davis, *The Cavaliers and Roundheads of Barbados, 1650–1652* (Georgetown, British Guiana: Argosy Press, 1887); Vincent Harlow, *A History of Barbados* (Oxford: Clarendon Press, 1926); James Williamson, *The Caribbee Islands under the Proprietary Patents* (Oxford: Oxford University Press, 1926).

3. Davis, *Cavaliers and Roundheads*, p. v.

4. Charles Andrews, *The Colonial Background of the American Revolution*, rev. ed. (New Haven: Yale University Press, 1931), pp. 122–28, and *The Colonial Period in American History*, vol. 4 (New Haven: Yale University Press, 1938), pp. 5–7.

5. Stephen Saunders Webb, *The Governors General: The English Army and the Definition of the Empire, 1569–1681* (Chapel Hill: University of North Carolina Press, Institute of Early American History and Culture, 1979).

6. Ibid., p. xvii.

CHAPTER 1

1. Alexander Brown, *The First Republic in America* (New York: Houghton Mifflin, 1898), pp. 256, 277, 326, 368, 370, 600; Wesley Frank Craven, *Dissolution of the Virginia Company* (New York: Oxford University Press, 1932), pp. 129–38; Vincent Harlow, *A History of Barbados, 1625–1685* (Oxford: Clarendon Press, 1926), p. 3.

2. G. T. Barton, *Prehistory of Barbados* (Bridgetown: Advocate Press, 1953).

3. Otis Starkey, *The Economic Geography of Barbados* (New York: Columbia University Press, 1939), pp. 45–49.

4. *Colonising Expeditions to the West Indies and Guiana, 1623–1667*, ed. Vincent T. Harlow (London: Hakluyt Society, 1925), p. 68.

5. *Verney Papers*, ed. John Bruce (London: Camden Society, 1853), p. 195.

6. *Colonising Expeditions*, p. 65.

7. Gary Puckrein, "Climate, Health and Black Labor in the English Americas," *Journal of American Studies* 13 (1979): 182–86, and "Humoralism and Social Development in Colonial America," *Journal of the American Medical Association* 245 (1981): 1755–57; Richard Leader to John Winthrop, Jr., August 14, 1660, quoted in *Massachusetts Historical Society Proceedings*, 2d ser., 3 (1886–87): 195–97; "A Swiss Medical Doctor's Description of Barbados in 1661: The Account of Felix Christian Spoeri," ed. and trans. Alexander Bunkel and Jerome S. Handler, *Jour. Bar. Mus. Hist. Soc.* 33 (1969–70): 5–6.

8. See Appendix.

9. Thomas Trapham, *A Discourse of the State of Health in the Island of Jamaica* (London, 1679), pp. 10, 26; "His Majesty's Council for Virginia, A Declaration of the State of the Colony and Affairs in Virginia," in *Tracts and Other Papers Relating Principally to the Origins, Settlement and Progress of the Colonies in North America*, comp. Peter Force, vol. 3, no. 5 (Washington, D.C., 1844), pp. 3–4; John Hammond, *Leah and Rachel, or the Two Fruitful Sisters Virginia and Maryland*, ibid., no. 14, pp. 6–7, 10, 12.

10. Gary Puckrein, "Did Sir William Courteen Really Own Barbados?" *Huntington Library Quarterly* 44 (1981): 135–37.

11. Peter Laslett, *The World We Have Lost* (New York: Scribner's, 1965), pp. 2–26; *Households and Family in Past Times*, ed. Peter Laslett (Cambridge: Cambridge University Press, 1972), pp. 26–27.

12. Laslett, *The World We Have Lost*, pp. 1–2, 13.

13. Lawrence Stone, *The Family, Sex and Marriage in England, 1500–1800* (New York: Harper and Row, 1977), pp. 85–91, and *The Crisis of the Aristocracy, 1550–1641* (Oxford: Clarendon Press, 1965), pp. 199–201.

14. Stone, *The Family*, pp. 85–91; Mervyn James, *Family, Lineage, and Civil Society* (Oxford: Clarendon Press, 1974), pp. 32–33.

15. James, *Family, Lineage*, p. 38; R. H. Tawney, *The Agrarian Problem in the Sixteenth Century* (New York: Burt Franklin, reprinted 1961), pp. 343, 415–16, 418.

16. John Hales, *A Discourse of the commonweal of this realm of England*, ed. Mary Deward (Charlottesville, Va.: The Folger Shakespeare Library, 1969), pp. 85–94; Thomas Wilson, "The State of England Anno Dom. 1600," ed. F. J. Fisher, *Camden Miscellany* 16 (1936): 38–41; James Harrington, *Commonwealth of Oceana*, ed. S. B. Liljegren (Heidelberg: Carl Winters, 1924), pp. 9–10, 16; *The English Works of Thomas Hobbes*, ed. Sir William Molesworth, vol. 3 (London, 1839), pp. 157–58, 319–22; John Locke, *Two Treaties of Government*, Introduction by Peter Laslett (Cambridge: Cambridge University Press, 1960), pp. 348–71.

17. *The Works of Francis Bacon*, ed. James Spedding, vol. 6 (London: Riverside Press, 1863), pp. 126, 179–80.

18. William Harrison, *The Description of England*, ed. George Edeles (Ithaca: Cornell University Press, The Folger Shakespeare Library, 1968), p. 118; Mildred Campbell, *The English Yeomen under Elizabeth and the Early Stuarts* (New Haven: Yale University Press, 1942), passim; Tawney, *The Agrarian Problem*, pp. 20–21, 34, 40, 343–47.

19. Christopher Hill, "The Poor and the People in 17th-Century England" (Social

History Group, Rutgers University, 1981); C. S. L. Davies, "Slavery and Protector Somerset: The Vagrancy Act of 1547," *Economic History Review*, 2d ser., 19 (1966): 533–49; Peter Slack, "Vagrants and Vagrancy in England, 1598–1644," *Economic History Review*, 2d ser., 27 (1974): 360–79; A. L. Beier, "Vagrants and Social Order in Elizabethan England," *Past and Present* 64 (1974):3–29.

20. Joyce Appleby, *Economic Thought and Ideology in Seventeenth-Century England* (Princeton: Princeton University Press, 1978); Thomas Wilson, *A Discourse upon Usury*, Introduction by R. H. Tawney (New York: Augustus M. Kelly, 1963), pp. 1–172.

21. *Winthrop Papers*, vol. 1 (Cambridge: Massachusetts Historical Society, 1931), pp. 356–57, 361–62.

22. Ibid., vol. 2, p. 67.

23. Ibid., vol. 5, p. 254.

24. Thomas Verney to Sir Edmund Verney, February 10, 1638, *Verney Papers*, p. 192.

25. *Memorials of the Great Civil War in England from 1646 to 1652*, ed. Henry Cary, vol. 2 (London: H. Colburn, 1842), p. 313.

26. "T. Walduck's Letters from Barbados, 1710," *Jour. Bar. Mus. Hist. Soc.* 15 (1947–48): 49–50.

27. Jerome S. Handler, "Father Antoine Biet's Visit to Barbados in 1654," *Jour. Bar. Mus. Hist. Soc.* 32 (1967): 67.

28. Morgan Godwin, *The Negro's and Indian's Advocate* (London, 1680), p. 81.

29. A good introduction to the topography of West Africa and its role in social development can be found in A. L. Mabogunje, "The Land and Peoples of West Africa," in *History of West Africa*, ed. J. F. Ade Ajayi and Michael Crowder, vol. 1 (New York: Columbia University Press, 1972), pp. 1–32.

30. A useful exploration of African oral traditions is Jan Vansina, *The Children of Woot* (Madison: The University of Wisconsin Press, 1978).

31. Sidney Mintz and Richard Price, *An Anthropological Approach to the Afro-American Past: A Caribbean Perspective* (Philadelphia: Institute for the Study of Human Issues, 1976), pp. 4–11.

32. Monica Schuler, *Alas, Alas Kongo* (Baltimore: The Johns Hopkins University Press, 1980), pp. 33–34, and "Akan Slave Rebellions in the British Caribbean," *Savacou* 1 (1970): 8–31.

33. John S. Mbiti, *African Religions and Philosophy* (New York: Doubleday, 1970), pp. 130–42; *African Systems of Kinship and Marriage*, ed. A. R. Radcliffe-Brown and Daryll Forde (London: International African Institute, 1950), pp. 1–85; Basil Davidson, *The African Genius* (Boston: Little, Brown, 1969), pp. 23–106.

34. Mbiti, *African Religions*, pp. 130–42; Radcliffe-Brown and Forde, *African Kinship*, pp. 1–85.

35. Mbiti, *African Religions*, passim; Radcliffe-Brown and Forde, *African Kinship*, pp. 70–71; Davidson, *African Genius*, pp. 67–80.

36. Willy de Craemer, Jan Vansina, and Renée C. Fox, "Religious Movements in Central Africa: A Theoretical Study," *Comparative Studies in Society and History* 18 (1976): 458–75.

37. Davidson, *African Genius*, pp. 54–67.

38. Igor Kopytoff and Suzanne Miers, "African Slavery as an Institution of Marginality," in *Slavery in Africa*, ed. Suzanne Miers and Igor Kopytoff (Madison: The University of Wisconsin Press, 1977), pp. 3–78; Joseph Miller, "The Slave Trade in Congo

and Angola," in *The African Diaspora*, ed. Martin L. Kilson and Robert Rotberg (Cambridge: Harvard University Press, 1976), pp. 76–78.

CHAPTER 2

1. The Three Springs Plantation owned by Francis Skeete in 1638 was huge (over 4,500 acres). Governor Henry Hawley possessed an estate that was at least 1,000 acres, and the earl of Carlisle rented a 10,000-acre plantation to a consortium of London merchants. Recopied Deed Books, Bar. Archives, RB 3/1/7; Summary of the Contents of an Inquisition Regarding Captain Henry Hawley, Davis Papers, box 1, envelope 2; Gary Puckrein, "The Carlisle Papers," *Jour. Bar. Mus. Hist. Soc.* 35 (1978):309–10.

2. Instructions I would have observed by Mr. Richard Harwood in the management of my plantation, Rawlinson Ms. A. 348 (Bodleian Library), fol. 1; "Instructions for the management of Drax-Hall and the Irish-Hope plantations to Archibald Johnson," reprinted in William Belgrave, *A Treatise upon Husbandry or Planting* (Boston, 1755), p. 51; Mr. Perrymen to Samuel Crisp, March 6, 1693 (Cumbria Record Office) D/Lons/LBM1, fol. 53; Acts of Barbados, P.R.O., C.O. 30/2/16; Richard Hall, *Acts, Passed in the Island of Barbados from 1643 to 1762* (London, 1764), pp. 4–5.

3. Dunn, *Sugar and Slaves*, p. 97.

4. Thomas Povey to William Povey, April 3, 1658, Additional Ms. 11411 (British Library), fols. 129–31.

5. Anthony Salerno, "The Social Background of Seventeenth-Century Emigration to America," *Journal of British Studies* 19 (1979): 31–52. See also James Horn, "Servant Emigration to the Chesapeake in the Seventeenth Century," in *The Chesapeake in the Seventeenth Century*, ed. Thad Tate and David Ammerman (Chapel Hill: University of North Carolina Press, Institute of Early American History and Culture, 1979), pp. 51–95.

6. Dunn, *Sugar and Slaves*, p. 50.

7. Ibid., p. 326.

8. *Calendar of State Papers, Domestic Series, 1631–1633*, ed. John Bruce (London: HMSO, 1862), p. 433.

9. Bridenbaugh, *No Peace*, p. 18.

10. Thomas Verney to Sir Edmund Verney, February 10, 1638, *Verney Papers*, ed. John Bruce (London: Camden Society, 1853), p. 192.

11. Henry Winthrop to Emmanuel Downing, August 22, 1627, *Winthrop Papers*, vol. 1, pp. 356, 361–62.

12. "A Letter from Barbados in 1640," *Jour. Bar. Mus. Hist. Soc.* 27 (1960):124–25.

13. Richard Waterhouse, "England, the Caribbean, and the Settlement of Carolina," *Journal of American Studies* 19 (1975): 259–81.

14. *Colonising Expeditions to the West Indies and Guiana, 1623–1667*, ed. Vincent T. Harlow (London: Hakluyt Society, 1925), pp. 30–31.

15. Henry Winthrop to Emmanuel Downing, August 22, 1627, *Winthrop Papers*, vol. 1, pp. 356–57, 361–62.

16. *Colonising Expeditions to the West Indies*, p. 31.

17. [William Duke], *Some Memoirs of the First Settlement of Barbados* (Barbados, 1741), pp. 51–62.

18. An Account of Land Granted by Captain William Hawley, Hay Papers (S.R.O.).

19. Dunn, *Sugar and Slaves*, p. 75.

20. The Earl of Carlisle's Grant of 10,000 Acres of Land to the Merchants, Papers

of James Hay, First Earl of Carlisle, Relating to Lands in the West Indies Granted to the Merchant Adventurers, 1628–30 (Henry Huntington Library), HM 17; Recopied Deed Books, Bar. Archives, RB 3/1/6–9, 33, 141; RB 3/3/29.

21. F. C. Innes, "The Pre-Sugar Era of European Settlement in Barbados," *Journal of Caribbean History* 1 (1970): 8–9.

22. Ibid., pp. 10–11; for examples of tenancy see Recopied Deed Books, Bar. Archives, RB 3/1/171–72; RB 3/13/348; RB 3/3/623; RB 3/1/115.

23. Peter Hay to Sir James and Archibald Hay, October 9, 1639, Hay Papers (S.R.O.).

24. Herman J. Nieboer, *Slavery as an Industrial System: Ethnological Research* (The Hague: Martinus Nijhoff, 1910), passim; E. G. Wakefield, *A View of the Art of Colonization* (Oxford: Clarendon Press, reprinted 1914), pp. 165–81; Evsey D. Domar, "The Causes of Slavery or Serfdom: A Hypothesis," *Journal of Economic History* 30 (1970): 18–32. See also Sidney Mintz, *Caribbean Transformation* (Chicago: Aldine, 1974), pp. 64–66. For contemporary observations on the land/labor imbalance see Emmanuel Downing to John Winthrop [1645], "The Winthrop Papers," *Massachusetts Historical Society, Collections*, 4th ser., vol. 6 (Boston, 1863), p. 65. A critical assessment of the land/labor thesis can be found in Orlando Patterson, "The Structural Origins of Slavery: A Critique of the Nieboer-Domar Hypothesis from a Comparative Perspective," *Comparative Perspectives on Slavery in the New World Plantation Societies* (New York: The New York Academy of Sciences, 1977), pp. 12–34.

25. Bridenbaugh, *No Peace*, p. 32.

26. Henry Winthrop to Emmanuel Downing, August 27, 1627, *Winthrop Papers*, vol. 1, p. 357.

27. Gary Puckrein, "Climate, Health and Black Labor in the English Americas," *Journal of American Studies* 13 (1979): 179–93, and "Humoralism and Social Development in Colonial America," *Journal of the American Medical Association* 245, no. 17 (1981): 1755–57.

28. Nicholas Foster, *A Brief Relation of the Late Horrid Rebellion Acted on the Island of Barbados in the West Indies* (London, 1650), pp. 1–2.

29. Some surviving plantation inventories of the late 1630s and early 1640s can be found in Recopied Deed Books, Bar. Archives, RB 3/1/13, 14, 15, 289; Dunn, *Sugar and Slaves*, p. 54.

30. William Hilliard to Archibald Hay, December 18, 1637, Hay Papers (S.R.O.).

31. Summary of the Contents of an Inquisition Regarding Captain Henry Hawley, Davis Papers, box 1, envelope 2.

32. Dunn, *Sugar and Slaves*, p. 72.

33. Deposition of Captain Simon Gordon, July 25, 1660, P.R.O., C.O. 1/14/25; Gary Puckrein, "Did Sir William Courteen Really Own Barbados?" *Huntington Library Quarterly* 44 (1981): 136–37; Kenneth Andrews, *The Spanish Caribbean* (New Haven: Yale University Press, 1978), pp. 224–49.

34. P.R.O., Patent Roll, 3 Charles I, pt. 31, no. 15; an English translation can be found in P.R.O., C.O. 29/1.

35. P.R.O., Patent Roll, 3 Charles I, pt. 30, no. 1.

36. P.R.O., Patent Roll, 3 Charles I, pt. 6, no. 4.

37. The Earl of Carlisle's Grant of 10,000 Acres of Land to the Merchants, Papers of James Hay, First Earl of Carlisle, Relating to Lands in the West Indies Granted to the Merchant Adventurers, 1628–30 (Henry Huntington Library), HM 17.

38. The Earl of Carlisle's Commission Granted to Charles Wolverston to be Governor of the Merchants' Plantation, Papers of James Hay, First Earl of Carlisle, Relating

to Lands in the West Indies Granted to the Merchant Adventurers, 1628–30 (Henry Huntington Library), HM 17: Puckrein, "Carlisle Papers," pp. 304–5.

39. Bill of Complaint of James, Earl of Carlisle, Marmaduke Rawden, William Perkins and Alexander Banister, Merchants, Captain William Deane of London, Gentlemen, September 9, 1629, P.R.O., C.O. Chancery Proceedings, ch. I, C60/no. 38 (i).

40. Deposition of Henry Powell, August 20, 1660, P.R.O., C.O. 1/14/39.

41. Appointment of Assistants to Barbados, September 4, 1628, *Winthrop Papers*, vol. 1, p. 405.

42. Ibid.

43. Ibid., p. 361; Bill of Complaint of James, Earl of Carlisle, Marmaduke Rawden, William Perkins and Alexander Banister, Merchants, Captain William Deane of London, Gentlemen, September 9, 1629, P.R.O., C.O. Chancery Proceedings, ch. I, C60/no. 38 (i). Deposition of Thomas Parris, August 20, 1660, P.R.O., C.O. 1/14/31.

44. Bill of Complaint of James, Earl of Carlisle, Marmaduke Rawden, William Perkins and Alexander Banister, Merchants, Captain William Deane of London, Gentlemen, September 9, 1629, P.R.O., C.O. Chancery Proceedings, ch. I, C60/no. 38 (i).

45. Bill Complaint of James, Earl of Carlisle, Marmaduke Rawden, William Perkins and Alexander Banister, Merchants, Captain William Deane of London, Gentlemen, September 9, 1629, P.R.O., C.O. Chancery Proceedings, ch. I, C60/no. 38 (i).

46. Deposition of Thomas Parris, August 20, 1660, P.R.O., C.O. 1/14/31; The Answer of Henry Powell One of the Defendants to the Bill of Complaint of the Right Honorable James, Earl of Carlisle, October 28, 1629, P.R.O., Chancery Proceedings, ch. I, C58/no. 4; Puckrein, "Sir William Courteen," p. 141.

47. Papers relating to the early history of Barbados, Davis Papers, box 1, envelope 1; Captain John Fincham to [Secretary Dorchester], October 12, 1629, P.R.O., C.O. 1/5/29.

48. Papers relating to the early history of Barbados, Davis Papers, box 1, envelope 1; The case of Sir William Tufton, Davis Papers, box 1, envelope 11.

49. Papers relating to the early history of Barbados, Davis Papers, box 1, envelope 1.

50. Captain John Fincham, A Description of Barbados, Coke Papers (Melbourne Hall, Derby).

CHAPTER 3

1. "The Voyage of Sir Henry Colt," *Colonising Expeditions to the West Indies and Guiana, 1623–1667*, ed. Vincent T. Harlow (London: Hakluyt Society, 1925), pp. 66–67.

2. Edmund Morgan, *American Slavery, American Freedom* (New York: W. W. Norton, 1975), pp. 108–11; Joyce Lorimer, "The English Contraband Tobacco Trade from Trinidad and Guiana, 1590–1617," in *The Westward Enterprise*, ed. K. R. Andrews, N. P. Canny, and P. E. H. Hair (Detroit: Wayne State University Press, 1979), pp. 124–50; Bridenbaugh, *No Peace*, pp. 52–53; F. C. Innes, "The Pre-Sugar Era of European Settlement in Barbados," *Journal of Caribbean History* 1 (1970): 13–14.

3. *Winthrop Papers*, vol. 2, p. 67.

4. Innes, "The Pre-Sugar Era," pp. 14–17; Robert Carlyle Batie, "Why Sugar? Economic Cycles and the Changing of Staples on the English and French Antilles, 1624–54," *Journal of Caribbean History* 8 (1976): 4–13.

5. Conveyances and Related Papers Concerning the Revenues of the Caribbee Islands Used by James, Earl of Carlisle, to Pay Debts Owing to Ann Henshaw (Hertfordshire Record Office, Hertford, England), 12544–52, 12621–23.

6. J. H. Bennett, "Peter Hay, Proprietary Agent in Barbados," *Jamaican Historical Review* 5 (1965): 13.

7. Ibid., p. 18.

8. Ibid.

9. Ibid., pp. 18–19; Demands of Mr. Peter Hay, Esquire to the Governor and Council, Hay Papers (S.R.O.).

10. Bennett, "Peter Hay," pp. 19–20.

11. Ibid., pp. 2–21; James Williamson, *The Caribbee Islands under the Proprietary Patents* (Oxford: Oxford University Press, 1926), pp. 188–89.

12. Peter Hay to James and Archibald Hay, May 24, 1639, Hay Papers (S.R.O.).

13. Copy of the Earl of Warwick's letter to the Governor and Council of Barbados [1639], Hay Papers (S.R.O.).

14. Bennett, "Peter Hay," p. 21; Peter Hay to James and Archibald Hay, May 24, 1639, Hay Papers (S.R.O.).

15. Bennett, "Peter Hay," p. 21; Peter Hay to James and Archibald Hay, May 24, 1639, Hay Papers (S.R.O.).

16. Thomas Verney to Sir Edmund Verney, February 10, 1638, *Verney Papers*, ed. John Bruce (London: Camden Society, 1853), p. 193.

17. Notes Relating to the Settlement of Barbados and St. Christopher Made at a Meeting of a Committee of the House of Commons, March 15–16, 1646/47, and April 19, 1647, Rawlinson Ms. 94 (Bodleian Library), fol. 33.

18. Ibid.

19. Bennett, "Peter Hay," p. 21; Peter Hay to James and Archibald Hay, May 24, 1639 (S.R.O.).

20. Bennett, "Peter Hay," p. 21; Williamson, *Caribbee Islands*, p. 136; Vincent Harlow, *A History of Barbados, 1625–1685* (Oxford: Oxford University Press, 1926), p. 18; Archibald to Peter Hay, May 27, 1639; Archibald Hay to Governor and Council of Barbados, May 27, 1639; Archibald Hay to Mr. Hilliard, January 30, 1639/40, Hay Papers (S.R.O.).

21. King to Governor and Council of Barbados, March 16, 1639, P.R.O., C.O. 1/10/13.

22. Copy of Commission to Captain Henry Hawley to Negotiate for the Cessation of the Planting of Tobacco in Barbados, March 27, 1639, Bankes Papers (Bodleian Library), 41/64.

23. G. H. Hawtayne, "Records of Old Barbados," *Timehri*, n.s., 10 (1896): 94, 96; An Inventory of the Papers and Writings Sent to the Right Earl of Carlisle, Hay Papers (S.R.O.); Copy of Sergeant Major Henry Huncks His Letter to My Lord the Earl of Carlisle, July 11, 1639, P.R.O., C.O. 1/10/75; Articles Against Captain Henry Hawley, P.R.O., C.O. 1/10/78.

24. Copy of Sergeant Major Henry Huncks His Letter to My Lord the Earl of Carlisle, July 11, 1639, P.R.O., C.O. 1/10/75; Articles Against Captain Henry Hawley, P.R.O., C.O. 1/10/78.

25. Copy of Sergeant Major Henry Huncks His Letter to My Lord the Earl of Carlisle, July 11, 1639, P.R.O., C.O. 1/10/75; Articles Against Captain Henry Hawley, P.R.O., C.O. 1/10/78.

26. Commission to Henry Ashton, Peter Hay, and others, *C.S.P.C. (1574–1660)*, p. 305.

27. The Commissioners to the King, June 30, 1640. P.R.O., C.O. 1/10/70.

28. William Powry to Archibald Hay, November 18, 1640. Hay Papers (S.R.O.).

29. William Powry to Archibald Hay, October 5, 1640, Hay Papers (S.R.O.).

30. Bennett, "Peter Hay," p. 24.

31. Ibid., p. 25.

32. Ibid., pp. 25–26.

33. Charges Against Peter Hay, February 1640/41, Davis Papers, box 7, envelope 12.

34. Bennett, "Peter Hay," pp. 26–27.

35. Ibid., p. 27.

36. Ibid., pp. 27–28; Peter Hay to [Archibald and James Hay], April 13, 1641, Hay Papers (S.R.O.).

37. Peter Hay to Trustees, February 23 and April 13, 1641, Hay Papers (S.R.O.).

38. Bennett, "Peter Hay," p. 28; Recopied Deed Books, Bar. Archives, RB 3/1/55.

39. Bennett, "Peter Hay," pp. 28–29; Peter Hay to Archibald Hay, June 22, 1641; Charges Against Peter Hay, Davis Papers (Roy. Com. Soc.), box 1, envelope 12.

40. J. H. Bennett, "The English Caribbees in the Period of the Civil War, 1642–1646," *William and Mary Quarterly*, 3d ser., 24 (1967): 368.

41. John Smith, *The General Historie of Virginia, New England and the Summer Isles*, vol. 2 (Glasgow: James MacLehose and Sons, 1907), p. 197; Captain John Fincham, A Description of the Barbados, Coke Papers (Melbourne Hall, Derby).

42. Father Andrew White, "A Brief Narrative of the Voyage unto Maryland," *Narratives of Early Maryland*, ed. Clayton Hall, in *Original Narratives of Early American History* (New York: Scribner's, 1910), p. 34.

43. Innes, "The Pre-Sugar Era," pp. 15–16.

44. Harlow, "The Voyage of Sir Henry Colt," p. 74.

45. Ibid., p. 69.

46. Father White, "Voyage unto Maryland," p. 35.

47. Archibald Hay to Peter Hay, October 10, 1637, Hay Papers (S.R.O.).

48. Peter Hay to James and Archibald Hay, April 13, 1638, Hay Papers (S.R.O.).

49. Daniel Fletcher to Archibald Hay, June 25, 1640, Hay Papers (S.R.O.).

50. John Scott, "The Description of Barbados," Sloane Ms. 2992 (British Library), fol. 55.

51. Peter Hay to Archibald and James Hay, May 30, 1637, Hay Papers (S.R.O.); Russell Menard, "A Note on Chesapeake Tobacco Prices, 1618–1660," *Virginia Magazine of History and Biography* 84 (1976): 405.

52. Many years later the colonists retained a deep gratitude to the Dutch for their aid in this time of crisis. See "A Declaration of Lord Willoughby and the Legislature of the Island of Barbados Against the British Parliament," quoted in Robert Schomburgk, *The History of Barbados* (London: Frank Cass, reprinted in 1971), p. 707.

53. David Watts, "A Seventeenth-Century Experiment in Barbadian Agricultural Improvisation," *Jour. Bar. Mus. Hist. Soc.* 34 (1972): 58.

54. Tobacco Entered in the Port of London, Additional Ms. 35865 (British Library), fol. 241.

CHAPTER 4

1. Richard Ligon, *A True and Exact History of Barbados* (London, 1657), p. 96.

2. Ibid., p. 37; *Winthrop's Journal*, ed. James Hosmer, in *Original Narratives of Early American History*, vol. 2 (New York: Scribner's, 1908), p. 328.

3. J. H. Lefroy, *Memorials of the Discovery and Early Settlement of the Bermudas or Somer*

Islands, vol. 1 (London: Longmans, Green, 1877), pp. 136, 154; Minutes of the Providence Island Company, P.R.O., C.O. 124/1/16–17.

4. James and Archibald Hay to Peter Hay, November 8, 1638, Hay Papers (S.R.O.); Robert Carlyle Batie, "Why Sugar? Economic Cycles and the Changing of Staples on the English and French Antilles, 1624–54," *Journal of Caribbean History* 8 (1976):1–41 provides a general overview of the change to sugar cultivation in the Caribbean, though his explanation for the change differs from the one presented here.

5. Ligon, *Barbados*, p. 85.

6. Thomas Robinson to Chappel, September 24, 1643, Hay Papers (S.R.O.).

7. Ligon, *Barbados*, p. 85.

8. Bridenbaugh, *No Peace*, p. 81.

9. Ligon, *Barbados*, p. 85.

10. A Coppie Journall Entries Made in the Custom House of Barbados Beginning August the 10th, 1664 and Ending August the 10th, 1665, Ms. Eng. hist. b. 122 (Bodleian Library); A Coppie Journall of Entries Made in the Custom House of Barbados 1665–1667, M. 1480 (Hispanic Society of America, New York City).

11. Vincent T. Harlow, *A History of Barbados, 1625–1685* (Oxford: Clarendon Press, 1962), p. 146; see also below, pages 139–140.

12. *Winthrop Papers*, vol. 5, p. 254.

13. Ligon, *Barbados*, p. 96.

14. *Memorials of the Great Civil War in England from 1646 to 1652*, ed. Henry Cary, vol. 2 (London: H. Colburn, 1842), p. 313.

15. Bridenbaugh, *No Peace*, pp. 84–85; Dunn, *Sugar and Slaves*, p. 66.

16. Ligon, *Barbados*, p. 22.

17. William Powry to Archibald Hay, April 8, 1646, Hay Papers (S.R.O.).

18. Ligon, *Barbados*, p. 108.

19. Earl of Carlisle, A Declaration Manifesting His Care of and Affection to the Good and Welfare of the Inhabitants of the Island of Barbados, and of All Other People Under His Government (John Carter Brown Library, Brown University).

20. [William Duke], *Some Memoirs of the First Settlement of the Island of Barbados* (Barbados, 1741), p. 18; Richard Pares, "Merchants and Planters," *Economic History Review Supplements*, no. 4 (Cambridge: Cambridge University Press, 1960), p. 57, fn. 15.

21. Ibid.

22. Quoted in Harlow, *Barbados*, p. 43, fn. 2.

23. Richard Hall, *Acts, Passed in the Island of Barbados from 1643 to 1762* (London, 1764), p. 14, no. 11; p. 24, no. 24.

24. John Scott's Description of Barbados, Sloane Ms. (British Library), 3662.

25. Instructions I would have observed by Mr. Richard Harwood in the management of my plantation, Rawlinson Ms. A. 348 (Bodleian Library), fol. 6; Sir Jonathan Atkins to Lords of Trade and Plantations, July 14, 1676, *C.S.P.C. (1675–1676)*, p. 420; Jerome Handler and Ron Shelby, "A Seventeenth-Century Commentary on Labor and Military Problems in Barbados," *Jour. Bar. Mus. Hist. Soc.* 34 (1973): 120; Gary Puckrein, "The Political Demography of Colonial Barbados" (Thirteenth Conference of the Association of Caribbean Historians, Pointe-à-Pitre, Guadeloupe, 1981).

26. *Winthrop Papers*, vol. 5, p. 172.

27. "A Letter From Barbados by the Way of Holland Concerning the Condition of Honest Men There, August 9, 1651," in *Colonising Expeditions to the West Indies and Guiana, 1623–1667*, ed. Vincent Harlow (London: Hakluyt Society, 1925), p. 51.

28. *Winthrop Papers*, vol. 5, p. 84.

29. *The Voyages of Captain Jackson (1642–1645)*, ed. Vincent Harlow, *Camden Miscellany*, 3d ser., 13 (1924): 15, 21–22.

30. Bridenbaugh, *No Peace*, p. 23.

31. Handler and Shelby, "Labor and Military Problems in Barbados," p. 119.

32. Alfred Chandler, "The Expansion of Barbados," *Jour. Bar. Mus. Hist. Soc.* 13 (1946): 106–36; Bridenbaugh, *No Peace*, pp. 198–229; Handler and Shelby, "Labor and Military Problems in Barbados," p. 118.

33. J. H. Bennett, "Cary Helyar, Merchant and Planter of the Seventeenth Century," *William and Mary Quarterly* 21 (1964): 59–60.

34. *Winthrop Papers*, vol. 5, p. 172.

35. For evidence of planters renting land see Recopied Deed Books, Bar. Archives, RB 3/13/348; RB 3/1/171–72, 191; RB 3/2/50, 544; RB 3/3/623.

36. Dunn, *Sugar and Slaves*, p. 97.

37. Bridenbaugh, *No Peace*, p. 131.

38. Agnes Whitson, *Constitutional Development of Jamaica, 1660–1729* (Manchester, England: Manchester University Press, 1929), pp. 29–30.

39. Ligon, *Barbados*, p. 96; P.R.O., S.P. 23/186/430.

40. Deposition of John Vincent and Lawrence Weltden, November 1648, London Mayor's Court Depositions, 1647–48 (Guildhall Record Office), box 2; Davis Papers, box 7, envelope 2; Recopied Deed Books, Bar. Archives, RB 3/1/536; RB 3/2/219–22; Lucas Papers (Barbados Free Public Library), vol. 5, pp. 144–48, 241; *Proceedings and Debates of the British Parliament Respecting North America (1572–1688)*, ed. Leo Stock (Washington, D.C.: Carnegie Institution of Washington, 1924), pp. 18–189.

41. Recopied Deed Books, Bar. Archives, RB 3/3/526.

42. Kenneth Haley, *The First Earl of Shaftesbury* (Oxford: Clarendon Press, 1968), pp. 64, 300.

43. *Aspinwall Notarial Records* (Boston: Municipal Printing Office, 1903), passim; Bridenbaugh, *No Peace*, p. 96.

44. For a general overview of the life of poor whites in Barbados see Jill Sheppard, *The "Redlegs" of Barbados* (New York: KTO Press, 1977); Hilary Beckles, "Rebels and Reactionaries: The Political Responses of White Labourers to Planter-class Hegemony in Seventeenth-Century Barbados," *Journal of Caribbean History* 15 (1981): 1–19.

45. For an alternative view of the origins of slavery in Barbados see Hilary Beckles, "The Economic Origins of Black Slavery in the British West Indies, 1640–1880: A Tentative Analysis of the Barbados Model," *Journal of Caribbean History* 16 (1982): 36–56.

46. Philip Curtin, *Economic Change in Precolonial Africa* (Madison: The University of Wisconsin Press, 1975), pp. 154–96; A. F. C. Ryder, *Benin and the Europeans, 1485–1897* (New York: Humanities Press, 1969), pp. 27–28.

47. General treatments of the Atlantic slave trade can be found in Walter Rodney, *A History of the Upper Guinea Coast, 1545–1800* (Oxford: Clarendon Press, 1970), pp. 95–121; David Northrup, *Trade without Rulers* (Oxford: Clarendon Press, 1978), pp. 50–84; David Birmingham, *Trade and Conflict in Angola* (Oxford: Clarendon Press, 1966), pp. 21–161; Philip Curtin, "The Atlantic Slave Trade, 1600–1800," in *History of West Africa*, ed. J. F. A. Ajayi and Michael Crowder, vol. 1 (New York: Columbia University Press, 1972), pp. 240–68; Joseph Miller, "The Slave Trade in Congo and Angola," in *The African Diaspora*, ed. Martin L. Kilson and Robert Rotberg (Cambridge: Harvard Univer-

sity Press, 1976), pp. 75–113; Ryder, *Benin and the Europeans*, pp. 42–75; Kwame Yeboa Daaku, *Trade and Politics on the Gold Coast* (Oxford: Clarendon Press, 1970), pp. 21–47; Basil Davidson, *The African Slave Trade* (Boston: Little, Brown, 1961); James A. Rawley, *The Transatlantic Slave Trade* (New York: W. W. Norton, 1981).

48. Rodney, *Upper Guinea*, pp. 102–3; Ryder, *Benin and the Europeans*, pp. 24–31; K. G. Davies, *The Royal African Company* (New York: Longmans, Green, 1957), pp. 226–27; Birmingham, *Trade and Conflict*, pp. 21–161; Miller, "Slave Trade in Congo and Angola," pp. 77–102.

49. Rawley, *Transatlantic Slave Trade*, pp. 151–52.

50. J. W. Blake, "The Farm of the Guinea Trade in 1631," in *Essays in British and Irish History*, ed. H. A. Cronne et al. (London: F. Muller, 1949), pp. 86–103; Alexander Brown, *The First Republic in America* (New York: Houghton Mifflin, 1898), pp. 288–89; *Documents Illustrative of the Slave Trade in America*, ed. Elizabeth Donnan, vol. 1 (Washington, D.C.: Carnegie Institution of Washington, 1930), pp. 78–79.

51. Lefroy, *Memorials of the Bermudas*, p.127.

52. Brown, *First Republic*, pp. 256, 277; Wesley Frank Craven, *Dissolution of the Virginia Company* (New York: Oxford University Press, 1932), pp. 129–38.

53. Russell Menard, "From Servants to Slaves: The Transformation of the Chesapeake Labor System," *Southern Studies* 16 (1977): 355–90.

54. Rawley, *Transatlantic Slave Trade*, p. 152.

55. Minutes of the Providence Island Company, P.R.O., C.O. 124/2/68–70; Arthur Percival Newton, *The Colonising Activities of the English Puritans* (New Haven: Yale University Press, 1914), pp. 35, 258.

56. James Browne to James and Archibald Hay, January 17, 1642, Hay Papers (S.R.O.).

57. *Winthrop Papers*, vol. 5, p. 43.

58. R. Porter, "The Crispe Family and the African Trade in the Seventeenth Century," *Journal of African History* 9 (1968): 57–68; Blake, "Farm of the Guinea Trade," pp. 87–94.

59. Pares, "Merchants and Planters," p. 75, fn. 33; *Documents Illustrative of the Slave Trade*, pp. 80, 82, 128–33; Deposition of Nicholas Crispe, Samuel Crispe, and John Wood, February 28, 1642/43, Davis Papers, box 7, envelope 2; Recopied Deed Books, Bar. Archives, RB 3/1/203.

60. [Henry Ashton] to Earl of Carlisle, n.d., Hay Papers (S.R.O.).

61. *Deposition Book of Bristol (1643–1647)*, ed. Miss H. E. Nott, vol. 1 (Bristol: Bristol Record Office, 1935), pp. 37–38, 219–37.

62. *Documents Illustrative of the Slave Trade*, pp. 1–17, 73–78.

63. Philip Curtin, *The Atlantic Slave Trade: A Census* (Madison: The University of Wisconsin Press, 1969), pp. 54–55.

64. An Estimate of the Barbados and of the Now Inhabitants There, Egerton Ms. 2395 (British Library), fol. 625.

CHAPTER 5

1. An interesting comparison of African and English magical beliefs can be drawn from Keith Thomas, *Religion and the Decline of Magic* (New York: Scribner's, 1971), and John S. Mbiti, *African Religions and Philosophy* (New York: Doubleday, 1970). See also

"T. Walduck's Letters from Barbados, 1710," *Jour. Bar. Mus. Hist. Soc.* 15 (1947–48): 148–49; Hutsen Crisp to Samuel Crisp, May 31, 1693 (Cumbria Record Office) D/Lons/LBM1, fol. 63.

2. Jay Mandle, *The Roots of Black Poverty* (Durham, N.C.: Duke University Press, 1978), pp. 3–15.

3. Bridenbaugh, *No Peace*, p. 119.

4. Jerome S. Handler, "Father Antoine Biet's Visit to Barbados in 1654," *Jour. Bar. Mus. Hist. Soc.* 32 (1967): 65.

5. Thomas Trapham, *A Discourse of the State of Health in the Island of Jamaica* (London, 1679), p. 26.

6. Frank Pitman, "Slavery on the British West India Plantations in the Eighteenth Century," *Journal of Negro History* 11 (1926): 584–668.

7. Philip Curtin, *The Atlantic Slave Trade: A Census* (Madison: The University of Wisconsin Press, 1969), p. 277.

8. Monica Schuler *Alas, Alas Kongo* (Baltimore: The Johns Hopkins University Press, 1980), pp. 33–34, and "Akan Slave Rebellions in the British Caribbean," *Savacou* 1 (1970): 8–31; see below, pages 162–164.

9. Elsa Goveia, "The West Indian Slaves Laws of the Eighteenth Century," in *Slavery in the New World*, ed. Laura Foner and Eugene D. Genovese (Englewood Cliffs, N.J.: Prentice-Hall, 1969), pp. 118–19.

10. [William Duke], *Some Memoirs of the First Settlement of the Island of Barbados* (Barbados, 1741), p. 9.

11. Morgan Godwin, *The Negro's and Indian's Advocate* (London, 1680), p. 81.

12. Richard Ligon, *A True and Exact History of the Island of Barbados* (London, 1657), p. 50; Marcus Jernegan, *Laboring and Dependent Classes in Colonial America, 1607–1783* (Westport, Conn.: Greenwood Press, reprinted 1980), pp. 24–44; Charles Verlinden, *The Beginnings of Modern Colonization*, trans. Yvonne Freccro (Ithaca: Cornell University Press, 1970), pp. 79–97.

13. *Great News From the Barbados. Or a True and Faithful Account of the Grand Conspiracy of the Negroes Against the English. And the Happy Discovery of the Same with the Number of Those that were Buried Alive, Beheaded, and Otherwise Executed for their Horrid Crimes. With a Short Description of that Plantation* (London, 1676), p. 6.

14. Ligon, *Barbados*, p. 29; Christopher Hill, "The Poor and the People in Seventeenth-Century England" (Social History Group, Rutgers University, 1981).

15. John Locke described the master/slave relationship as a state of war. John Locke, *Two Treatises of Government*, Introduction by Peter Laslett (Cambridge: Cambridge University Press, 1960), pp. 402–15.

16. Instructions I would have observed by Mr. Richard Harwood in the management of my plantation, Rawlinson Ms. A. 348 (Bodleian Library), fol. 7.

17. Ibid.

18. Ibid.

19. Handler, "Father Antoine Biet's Visit," p. 67.

20. "Father Labat's Visit to Barbados in 1700," trans. Neville Connell, *Jour. Bar. Mus. Hist. Soc.* 24 (1957): 169.

21. Richard Hall, *Acts, Passed in the Island of Barbados from 1643 to 1762* (London, 1764), pp. 4–5.

22. "Instructions for the Management of Drax-Hall and the Irish-Hope Plantations,"

reprinted in William Belgraves, *A Treatise upon Husbandry or Planting* (Boston, 1755), p. 52.

23. Godwin, *The Negro's Advocate*, p. 101.
24. Ligon, *Barbados*, p. 50.
25. Ibid.
26. Godwin, *The Negro's Advocate*, p. 61.
27. "Instructions for the Management of Drax-Hall and the Irish-Hope Plantations," pp. 55–56.
28. Instructions I would have observed by Mr. Richard Harwood in the management of my plantation, Rawlinson Ms. A. 348 (Bodleian Library), fol. 14.
29. Ibid.
30. Ligon, *Barbados*, p. 48.
31. Ibid., p. 47.
32. Ibid.
33. Instructions I would have observed by Mr. Richard Harwood in the management of my plantation, Rawlinson Ms. A. 348 (Bodleian Library), fol. 14.
34. Ligon, *Barbados*, pp. 46, 49–50, 52–54.
35. For a more detailed discussion of the gang system see Pitman, "Slavery on the British West India Plantations," pp. 595–609.
36. "Instructions for the Management of Drax-Hall and the Irish-Hope Plantations," pp. 65–66. The best treatment of the sugarmaking process in the seventeenth century is Ward Barrell, "Caribbean Sugar-Production in the Seventeenth and Eighteenth Centuries," in *Merchants and Scholars*, ed. John Parker (Minneapolis: University of Minnesota Press, 1965), pp. 147–70; see also Noel Deerr, *The History of Sugar*, vol. 2 (London: Chapman and Hall, 1949–50).
37. "Instructions for the Management of Drax-Hall and the Irish-Hope Plantations," p. 66.
38. Ibid.
39. Ibid.
40. The question of African retentions and the acculturation of Negroes in the New World has been widely discussed, and various schools of thought have emerged. At one extreme is the anthropologist Melville Herskovits, who argues for the survival of African culture in the Americas. At the opposite end is the sociologist E. Franklin Frazier, who maintains that African peoples were completely deculturated by the slave experience. Sidney Mintz and Richard Price have recently begun to construct a via media between these two extremes. In their view, Afro-American culture is to be seen as a New World development, but they recognize that it was deeply influenced by the traditional African world. The position taken here is more in keeping with the work of Mintz and Price. See Melville Herskovits, *The Myth of the Negro Past* (New York: Harper and Brothers, 1941); Franklin Frazier, *The Negro Family in the United States* (Chicago: University of Chicago Press, 1939), passim; Sidney W. Mintz and Richard Price, *An Anthropological Approach to the Afro-American Past: A Caribbean Perspective, Institute for the Study of Human Issues, Occasional Papers in Social Change*, no. 2 (Philadelphia, 1976), passim.
41. Ligon, *Barbados*, p. 46.
42. Minutes of the Barbadian Council, September 2, 1657, P.R.O., Round Room.
43. E. P. Thompson, "The Moral Economy of the English Crowd in the Eighteenth

Century," *Past and Present*, no. 5 (1971): 78–79, and "Patrician Society, Plebeian Culture," *Journal of Social History* 7 (1974); John Walter and Keith Wrightson, "Dearth and Social Order in Early Modern England," *Past and Present*, no. 71 (1976): 22–23; Peter Clark, "Popular Protest and Disturbance in Kent, 1558–1640," *Economic History Review* 29 (1976): 371–73; *The Works of Francis Bacon*, ed. James Spedding, vol. 6 (London: Riverside Press, 1863), p. 179.

44. *Tudor Economic Documents*, ed. R. H. Tawney and Eileen Power, vol. 3 (New York: Barnes and Noble, 1963), p. 313.

45. Acts of Barbados, P.R.O., C.O. 30/2/16–17.

46. Acts of Barbados, P.R.O., C.O. 30/2/110.

47. Minutes of the Assembly, Bar. Archives, HA 3/2/273.

48. Acts of Barbados, P.R.O., C.O. 30/2/108–11.

49. A perusal of Barbadian slave codes will prove the point. See Acts of Barbados, P.R.O., C.O. 30/2/16–17, 114–28; Hall, *Acts of Barbados*, p. 113.

50. Governor Sir Jonathan Atkins to Secretary Sir Joseph Williamson, October 3, 1675, *C.S.P.C. (1675–1676)*, pp. 294–95; Minutes of the Barbadian Council, February 16, 1686, *C.S.P.C. (1685–1688)*, p. 155; Minutes of the Barbadian Council, Transcript, 1689–96, Bar. Archives, fol. 274; *Great News from the Barbados*, passim.

51. See, for example, Minutes of the Council, February 16, 1685/86, Bar. Archives.

CHAPTER 6

1. Several historians have noted the independence of Barbados in the 1640s. See particularly J. H. Bennett, "The English Caribbees in the Period of the Civil War, 1642–1646," *William and Mary Quarterly* 3d ser., 24 (1967): 359–77.

2. J. H. Lefroy, *Memorials of the Discovery and Early Settlement of the Bermudas or Somer Islands*, vol. 1 (London: Longmans, Green, 1877), pp. 389, 487; Arthur Percival Newton, *The Colonising Activities of the English Puritans* (New Haven: Yale University Press, 1914), pp. 93, 216, 219.

3. The Oath Fealty, 1640, Davis Papers, box 7, envelope 1; Carlisle's declaration, February 18, 1640/41, Davis Papers, box 1, envelope 12.

4. Carlisle's declaration, February 18, 1640/41, Davis Papers, box 1, envelope 12.

5. James Browne to Trustees, January 17, 1642, Hay Papers (S.R.O.); Bennett, "The English Caribbees," p. 368.

6. Archibald Hay to Captain Philip Bell, January 1644; Earl of Carlisle to Captain Philip Bell, March 10, 1642/43; Charles I to Edward Skipwith, November 28, 1643. Bell to Trustees, August 25, 1643, Hay Papers (S.R.O.).

7. Archibald Hay to Captain Philip Bell, January 1644, Hay Papers (S.R.O.).

8. Earl of Carlisle to Captain Philip Bell, March 10, 1642/43, Hay Papers (S.R.O.).

9. Earl of Carlisle to Captain Philip Bell, March 10, 1642/43, Hay Papers (S.R.O.).

10. Captain Philip Bell to Trustees [August 15, 1645], Hay Papers (S.R.O.).

11. James Browne to Trustees, August 15, 1643, Hay Papers (S.R.O.); Richard Hall, *Acts, Passed in the Island of Barbados from 1643 to 1762* (London, 1764), p. 459.

12. James Browne to Trustees, August 15, 1643, Hay Papers (S.R.O.).

13. Captain Philip Bell to Trustees [August 15, 1643], Hay Papers (S.R.O.).

14. Hall, *Acts of Barbados*, p. 6.

15. Ibid., p. 12.

16. James Browne to Trustees, September 15, 1643; Archibald Hay to Philip Bell, Council, and Assembly, April 17, 1644, Hay Papers (S.R.O.).

17. *Proceedings and Debates of the British Parliament Respecting North America (1572–1688)*, ed. Leo Stock, vol. 1 (Washington, D.C.: Carnegie Institution of Washington, 1924), pp. 145–49.

18. Ibid., p. 189.

19. Earl of Carlisle to Captain Philip Bell, January 5, 1644; Earl of Carlisle to Governor and Council of Barbados, January 1644, Hay Papers (S.R.O.).

20. A Proclamation to Give Assurances unto All His Majesty's Subjects in the Islands and Continent of America of His Majesty's Care Over Them and to Preserve Them in Their Due Obedience, November 24, 1643, Hay Papers (S.R.O.).

21. Earl of Carlisle to Captain Philip Bell, January 5, 1644; Earl of Carlisle to Governor and Council of Barbados, January 1644, Hay Papers (S.R.O.).

22. Earl of Carlisle to Governor and Council of Barbados, January 5, 1644, Hay Papers (S.R.O.).

23. Trustees to Governor, Council, and Assembly of Barbados, April 17, 1644, Hay Papers (S.R.O.); Bennett, "The English Caribbees," p. 371.

24. *Lords Journal*, vol. 5, pp. 280a, 334a, 335b, 338a, 351b, 362b, 366a, 514a; vol. 6, pp. 117b, 538b, 542a; William Powry to Archibald Hay, July 5, 1645; Philip Bell to Archibald Hay, July 21, 1645, Hay Papers (S.R.O.); Bennett, "English Caribbees," p. 373.

25. Bridenbaugh, *No Peace*, pp. 14, 16, 130, 382–83; *Winthrop Papers*, vol. 5, pp. 84, 254; Jerome S. Handler, "Father Antoine Biet's Visit to Barbados in 1654," *Jour. Bar. Mus. Hist. Soc.* 32 (1967): 68–69; Henry Wilkinson, *The Adventurers of Bermuda* (Oxford: Oxford University Press, 1933), pp. 260–61; Newton, *Colonising Activities of Puritans*, p. 279.

26. Handler, "Father Antoine Biet's Visit," p. 69.

27. James Parker to John Winthrop, 1646, *Winthrop Papers*, vol. 5, p. 84.

28. Philip Bell to Archibald Hay [August 25, 1643]; Archibald Hay to Governor, Council, and Assembly of Barbados, April 17, 1644; Archibald Hay to Captain Philip Bell, January 1644, Hay Papers (S.R.O.).

29. William Powry to Archibald Hay, July 5, 1645, Hay Papers (S.R.O.).

30. Richard Ligon, *A True and Exact History of Barbados* (London, 1657), pp. 57–58.

31. William Powry to Archibald Hay, July 5, 1645, Hay Papers (S.R.O.).

32. Governor, Council, and Assembly of Barbados to Earl of Warwick and Committee for Foreign Plantations [October 1646]; Stock, *Debates of Parliament*, p. 190.

33. William Powry to Archibald Hay, July 5, 1645, Hay Papers (S.R.O.).

34. William Powry to Archibald Hay, July 5, 1645, Hay Papers (S.R.O.).

35. William Powry to Archibald Hay, July 5, 1645; Philip Bell to Archibald Hay, July 21, 1645; William Powry to Archibald Hay, September 1645, Hay Papers (S.R.O.).

36. Stock, *Debates of Parliament*, p. 168.

37. Ibid., p. 176.

38. Earl of Warwick to Philip Bell, n.d., Stowe Ms. 184 (British Library), fol. 124.

39. Earl of Warwick to Philip Bell, n.d., Stowe Ms. 184 (British Library), fol. 124.

40. Earl of Wariwck to Philip Bell, n.d., Stowe Ms. 184 (British Library), fol. 124.

41. Earl of Warwick to Philip Bell, n.d., Stowe Ms. 184 (British Library), fol. 125.

42. Earl of Warwick to Philip Bell, n.d., Stowe Ms. 184 (British Library), fols. 125–26.

43. Earl of Warwick to Philip Bell, n.d., Stowe Ms. 184 (British Library), fols. 125–26.

44. Stock, *Debates of Parliament*, pp. 189–90.

45. James Parker to John Winthrop, 1646, *Winthrop Papers*, vol. 5, p. 84.

46. Governor, Council, and Assembly of Barbados to Earl of Warwick and Committee for Foreign Plantations [October 1646], Hay Papers (S.R.O.); Stock, *Debates of Parliament*, p. 190.

47. Stock, *Debates of Parliament*, p. 187.

48. Ibid., p. 188.

49. Ibid., p. 184.

50. Ibid., pp. 191–93.

51. *Dictionary of National Biography*, ed. Sir Leslie Stephen and Sir Sidney Lee, vol. 21 (Oxford: Oxford University Press, reprinted 1922), pp. 504–5.

52. The original deed of demise from the Earl of Carlisle to the Lord Willoughby for the Caribbee Islands for one and twenty years, February 17, 1645; Lord Willoughby's Letters Patent as Lieutenant General of the Caribbee Islands, February 13, 1645, Mss. G. 14, 15 (Trinity College, Dublin), fols. 5–14, 19–35.

53. Stock, *Debates of Parliament*, p. 193.

54. Ibid., p. 187.

55. Ibid., pp. 193–94.

56. Ibid., p. 196.

CHAPTER 7

1. James Drax was a kinsman of William Hilliard, who was a partner of Thomas Kendall, a prominent London merchant. Thomas Noel was the brother of Martin Noel; the latter was perhaps the most important London merchant in the Cromwellian period. John Bayes seems to have had ties with John Bradshaw, who was soon to be president of the Council of State.

2. John Bayes to Council of State, June 30, 1652, P.R.O., C.O. 1/11/59.

3. Thomason Tracts E. 777 (British Library), fol. 18.

4. A Declaration of Lord Willoughby and the Legislature of the Island of Barbados Against the British Parliament," quoted in Robert Schomburgk, *The History of Barbados* (London: Frank Cass, reprinted 1971), p. 706.

5. An Essay Evenly Discussing the Present Condition and Interest of Barbados. And Consideration for the Rendring the Island Peaceable to Itself and Useful to this Commonwealth without the Hazard and the Charge of Sending a Fleet. Edward E. Ayer Collection, Phillipps Ms. 9728 (Newberry Library, Chicago), fol. 4.

6. *Acts and Ordinances of the Interregnum, 1642–1660*, ed. C. H. Firth and R. S. Rait, vol. 1 (London: HMSO, 1911), p. 330.

7. Order of the Council of State, July 24, 1649, *C.S.P.C. (1574–1660)*, p. 330.

8. Order of the Council of State, March 15, 1649, and May 22, 1649, *C.S.P.C. (1574–1660)*, pp. 328–29.

9. George Beer, *Origin of the British Colonial System, 1578–1660* (New York: Macmillan, 1908), p. 362.

10. Bermuda was the possession of the Somers Island Company. A major force in the company was the earl of Warwick; as his presence would suggest, its members were

staunch supporters of Parliament. Bermudian colonists sided with Parliament by accepting the 1643 ordinance that named Warwick governor in chief of all the American colonies. Despite the widespread affection for the Westminster government, the religious question deeply divided Bermudians. In fact, throughout the English Civil War political control was fought over by a Presbyterian majority and a tiny but vocal sectarian minority. Ignoring the protestations of the Presbyterians, in 1646 Parliament granted religious freedom to the radicals. Angry at this decision, and determined to exile the sectarians—who, according to local rumor, were plotting a violent overthrow of the government—in 1648 the Presbyterians declared for Charles I and banished all radicals who would not swear to obey the colony's laws. Realizing that they could not protect themselves from retribution from the House of Commons, in the fall of 1649 the Bermudians sent an agent to Barbados seeking help against Parliament. J. H. Lefroy, *Memorials of the Discovery and Early Settlement of the Bermudas or Somer Islands*, vol. 1 (London: Longmans, Green, 1877), pp. 569–655; Henry Wilkinson, *The Adventurers of Bermuda* (Oxford: Oxford University Press, 1933), pp. 272–89; [Thomas Modyford], *A Brief Relation of the Beginning and Ending of the Troubles of the Barbados* (London, 1653), Egerton Ms. 2395 (British Library), fol. 208.

11. Modyford, *Brief Relation of the Troubles of Barbados*, fol. 210.

12. Ibid., fol. 208.

13. Richard Ligon, *A True and Exact History of Barbados* (London, 1657), pp. 45–46.

14. Modyford, *Brief Relation of the Troubles of Barbados*, fol. 209.

15. Nicholas Foster, *A Brief Relation of the Late Horrid Rebellion Acted on the Island of Barbados in the West Indies* (London, 1650), pp. 9–12.

16. Ibid., p. 18.

17. Modyford, *Brief Relation of the Troubles of Barbados*, fol. 209.

18. Foster, *Horrid Rebellion in Barbados*, pp. 24, 30.

19. Ibid., p. 37.

20. Ibid., pp. 37, 100–101.

21. Modyford, *Brief Relation of the Troubles of Barbados*, fol. 209.

22. Ibid.; Foster, *Horrid Rebellion in Barbados*, pp. 38, 98.

23. Foster, *Horrid Rebellion in Barbados*, pp. 40–43.

24. *Dictionary of National Biography*, ed. Sir Leslie Stephen and Sir Sidney Lee, vol. 21 (Oxford: Oxford University Press, reprinted 1922), pp. 504–5; *Memorials of the Great Civil War in England from 1646 to 1652*, ed. Henry Cary, vol. 2 (London: H. Colburn, 1842), p. 315; Bulstrode Whitelocke, *Memorials of the English Affair*, vol. 3 (Oxford: Oxford University Press, 1853), p. 215.

25. Modyford, *Brief Relation of the Troubles of Barbados*, fols. 209–10.

26. Ibid., fol. 210; Whitelocke, *Memorials*, p. 215; The Lord Willoughby's Patent to be Lieutenant General of the Caribbee Islands, February 26, 1647/48, P.R.O., C.O. 29/1/47–50; Foster, *Horrid Rebellion in Barbados*, p. 110.

27. Depositions of John Mountejoy, William Pipe, and William Pointer, P.R.O., H.C.A. 1/8/9–12.

28. Modyford, *Brief Relation of the Troubles of Barbados*, fol. 210; Foster, *Horrid Rebellion in Barbados*, p. 110.

29. Foster, *Horrid Rebellion in Barbados*, p. 51. For an analysis of the delinquent list compiled by the General Assembly see Gary Puckrein, "The Acquisitive Impulse: Plantation Society, Factions, and the Origins of the Barbadian Civil War (1627–1652)" (Ph.D. diss., Brown University, 1978), pp. 219–24.

30. To the Right Honorable Council of State the Humble Proposals of Several Barbadians, P.R.O., C.O. 1/11/25; Schomburgk, *Barbados*, p. 706.

31. Letter from Barbados by J.B., Thomason Tracts E. 777 (British Library), fol. 18.

32. Barbadian refugees appealed to the English audience by casting the rebels in a more Royalist light than was consistent with the truth. One promoter of Willoughby in England complained that the deported planters were making the island's stance appear to be as one "levelled directly against this commonwealth." English newspapers were filled with incendiary propaganda. Paraphrasing letters from Barbados that reached England shortly before the exiled colonists arrived, *The Perfect Passages* gave this completely inaccurate account of the Barbadian Civil War:

> the Lord Willoughby of Parham arrived there privately, and having a purpose to advance the service of the King of Scots, kept himself for a time secret and unknown, he gathered head, and then they proclaimed Charles the Second King of England, Scotland, France, and Ireland; and have imprisoned Lieut. Col. Drax, and Capt. Thomas Middleton and many others, upon no other ground but because they supposed them to be well affected to the Parliament of England, threatening to bring them to a speedy trial for their lives.

The *Impartial Scot* reported that Willoughby was stirring up the hearts of Barbadians to an "engagement for the restoration of their sovereignty" but that the "well-affected seek to oppose him." It was this type of inaccurate reporting that agitated one Englishman to prophesy that "sudden care will be taken for relief of the honest party there, and those hot spirits may have time to know that he who by himself or any other begin any dangerous enterprise, without foreseeing the end or issue thereof, may bring himself and others into peril." An Essay Evenly Discussing the Present Condition and Interest of Barbados. Edward E. Ayer Collection, Phillipps Ms. 9728 (Newberry Library, Chicago), fol. 2; *Perfect Passages of Everyday Intelligence from the Parliament*, Thomason Tracts E. 777 (British Library), fol. 24; *The Impartial Scot*, Thomason Tracts E. 777 (British Library), fol. 23.

33. Orders of the Committee of the Admiralty, August 27 and 30, 1650, *C.S.P.C. (1574–1660)*, p. 342.

34. An Essay Evenly Discussing the Present Condition and Interest of Barbados, fol. 6.

35. Ibid., fol. 7.

36. Ibid., fol. 1.

37. Ibid., fol. 4.

38. Ibid., fols. 8–12.

39. Orders of the Committee of the Admiralty, August 27, 1650, *C.S.P.C. (1574–1660)*, p. 342; *Acts and Ordinances*, pp. 425–29.

40. The four propositions were: (1) that all acts offensive to the Commonwealth be repealed; (2) that Charles II be renounced as king; (3) that the principal incendiaries in the conflict be banished (most likely the Walrond brothers); and (4) that all exiled persons in England be permitted to return to the island where they were to enjoy freedom of conscience and all other rights exercised by the free inhabitants of the colony. The salient and unspoken feature of the merchants' propositions was that the existing government was to continue in office. The Humble Desires of Divers Merchants of London and Planters Interested in the Island of Barbados, November, 20, 1650, P.R.O., C.O. 1/11/23.

41. John Bayes to [John Bradshaw], November 1650, P.R.O., C.O. 1/11/26.

42. Humble Proposals of Several Barbadians, P.R.O., C.O. 1/11/25.

43. *Proceedings and Debates of the British Parliament Respecting North America (1572–1688)*, ed. Leo Stock, vol. 1 (Washington, D.C.: Carnegie Institution of Washington, 1924), pp. 221–22; Council of State, Day's Proceedings, November 27, 1650, *Calendar of State Papers, Domestic Series, 1650*, ed. Mary Anne Everett Green (London: HMSO, 1876), p. 444; Commission Appointing Sir George Ayscue, Daniel Searle, and Captain Michael Pack Commissioners for Reducing the Island of Barbados; Warrant to Sir George Ayscue; Instructions to Sir George Ayscue, Daniel Searle, and Captain Michael Pack Commissioners for Reducing Barbados, *C.S.P.C. (1574–1660)*, pp. 349–50.

44. George Ayscue to Council of State, February 21, 1651/52, Tanner Ms. 55 (Bodleian Library), fol. 142b; Stock, *Debates of Parliament*, p. 221.

45. My Lord I Humbly Present to Your Hands my Thoughts Concerning the Island of Barbados, William Hilliard to John Bradshaw, P.R.O., C.O. 1/11/28.

46. Puckrein, "Acquisitive Impulse," pp. 219–24.

47. Modyford, *Brief Relation of the Troubles of Barbados*, fol. 210.

48. Ibid., fol. 211.

49. The Barbados Engagement, Tanner Ms. 56 (Bodleian Library), fol. 210b. Interestingly enough, the engagement omitted the king's name or any allegiance due the crown. When Samuel Farmer, a Royalist émigré, brought this treasonous omission to Willoughby's attention, the governor ordered him arrested. Petition of Samuel Farmer to the King and Council [November 15, 1665], P.R.O., C.O. 1/20/54.

50. An Act for the Borrowing of Goods for the Present Defence of the Island, April 3, 1651, P.R.O., C.O. 1/11/33.

51. A Declaration by the Lord Lieutenant General, and the Gentlemen of the Council and Assembly set Forth for the Satisfaction of all the Inhabitants of the Island of Barbados, June 11, 1651, P.R.O., C.O. 1/11/34, 43.

52. Schomburgk, *Barbados*, pp. 706–8.

53. Ligon, *Barbados*, p. 46.

54. Beauchamp Plantagenet, "A Description of the Province of New Albion," in *Tracts and Other Papers*, ed. Peter Force (Gloucester, Mass.: Peter Smith, reprinted 1963), no. 7, p. 5.

55. Jerome S. Handler and Ron Shelby, "A Seventeenth-Century Commentary on Labor and Military Problems in Barbados," *Jour. Bar. Mus. Hist. Soc.* 34 (1973):120.

56. Nicholas Blake to [Joseph Williamson], November 12, 1669, P.R.O., C.O. 1/24/94.

57. Thomas Modyford to Lord Bradshaw, February 1651/52, P.R.O., C.O. 1/11/41.

58. Handler and Shelby, "Labor and Military Problems in Barbados," p. 119.

59. An Act for the Prohibition of Landing Irish Persons, August 29, 1644; Richard Hall, *Acts, Passed in the Island of Barbados from 1643 to 1762* (London, 1764), p. 459.

60. Minutes of the Barbadian Council, September 1, 1657, P.R.O., Round Room.

61. Minutes of the Barbadian Council, January 15, 1655/56, P.R.O., Round Room.

62. Minutes of the Barbadian Council, November 6, 1655, P.R.O., Round Room.

63. Nicholas Darnell Davis, *The Cavaliers and Roundheads of Barbados, 1650–1652* (Georgetown, British Guiana: Argosy Press, 1887), p. 252.

64. Sir George Ayscue to Council, October 19, 1651, Tanner Ms. 55 (Bodleian Library), fol. 79; Sir George Ayscue to Council of State, October 31, 1651, Tanner Ms. 55 (Bodleian Library), fol. 85; The Weekly Intelligence of the Commonwealth, Thomason Tracts E. 655 (British Library).

65. Quoted in Davis, *Cavaliers and Roundheads*, pp. 217–18.

66. Sir George Ayscue to Council of State, February 18, 1652, Tanner Ms. 55 (British Library), fol. 141.

67. Sir George Ayscue to Council of State, February 18, 1652, Tanner Ms. 55 (British Library), fol. 141; George Ayscue to Lord Willoughby of Parham, December 14, 1651, P.R.O., C.O. 1/11/40.

68. Davis, *Cavaliers and Roundheads*, p. 252.

69. General Assembly to Sir George Ayscue, December 25, 1651, P.R.O., C.O. 1/11/40.

70. Sir George Ayscue to General Assembly, January 1, 1651/52, P.R.O., C.O. 1/11/40.

71. Sir George Ayscue to General Assembly, January 1, 1651/52, P.R.O., C.O. 1/11/40.

72. General Assembly to Sir George Ayscue, December 29, 1651, P.R.O., C.O. 1/11/40.

73. George Ayscue to Council of State, February 18, 1651/52, Tanner Ms. 55 (British Library), fol. 141.

74. Thomas Modyford to Lord Bradshaw, February 16, 1651/52, P.R.O., C.O. 1/11/41.

75. Modyford, *Brief Relation of the Troubles of Barbados*, fol. 210.

76. George Ayscue to Council of State, February 18, 1651/52, Tanner Ms. (British Library), fol. 141; Modyford, *Brief Relation of the Troubles of Barbados*, fol. 211.

77. Lord Willoughby to George Ayscue, January 9, 1651/52, P.R.O., C.O. 1/11/40.

CHAPTER 8

1. A printed copy of the articles of surrender can be found in Nicholas Darnell Davis, *The Cavaliers and Roundheads of Barbados, 1650–1652* (Georgetown, British Guiana: Argosy Press, 1887), pp. 251–55.

2. Ibid., p. 254.

3. Colonel Thomas Modyford to John Bradshaw, February 16, 1652, P.R.O., C.O. 1/11/117; Thomas Modyford to Oliver Lord Cromwell, Egerton Ms. 2395 (British Library), fol. 175.

4. Considerations to be prepared to the Council of State concerning the present settlement of Barbados [August 1652], P.R.O., C.O. 1/11/66.

5. *Proceedings and Debates of the British Parliament Respecting North America (1572–1688)*, ed. Leo Stock, vol. 1 (Washington, D.C.: Carnegie Institution of Washington, 1924), p. 229.

6. John Bayes to Council of State, June 30, 1652, P.R.O., C.O. 1/11/59.

7. John Bayes to Council of State, June 30, 1652, P.R.O., C.O. 1/11/59.

8. John Bayes to Committee for Foreign Affairs, P.R.O., C.O. 1/12/2; Daniel Searle to Council of State, September 19, 1653, P.R.O., C.O. 1/12/155.

9. The State of the Difference as it is Pressed Between the Merchants and the Planters in Relation to Free Trade and the Means of Reconciliation and General Satisfaction Proposed, Additional Ms. 11411 (British Library), fols. 3–19; John Bayes to Committee for Foreign Affairs [February 4, 1653], P.R.O., C.O. 1/12/2; J. E. Farnell, "The Navigation Act of 1651, the First Dutch War and the London Merchant Community," *Economic History Review*, 2d ser., 16 (1964): 440; Richard Sheridan, *Sugar and Slavery* (Bal-

timore: The Johns Hopkins University Press, 1973), pp. 40–43; Charles Andrews, *The Colonial Period in American History*, vol. 4 (New Haven: Yale University Press, 1938), pp. 35–36.

10. Farnell, "Navigation Act," pp. 439–54; Sheridan, *Sugar and Slavery*, pp. 40–45; Andrews, *Colonial Period*, passim.

11. John Bayes to Committee for Foreign Affairs [February 4, 1653], P.R.O., C.O. 1/12/2.

12. Minutes of a Committee for Foreign Affairs, November 17 and December 15, 1652, *C.S.P.C. (1574–1660)*, pp. 393–94.

13. Ibid., p. 396; Commission of Daniel Searle, Egerton Ms. 2395 (British Library), fol. 116.

14. John Bayes to Council of State, June 30, 1652, P.R.O., C.O. 1/11/59.

15. Daniel Searle to Council to State, August 28, 1653, P.R.O., C.O. 1/12/9.

16. Daniel Searle to Council of State, August 28, 1653, P.R.O., C.O. 1/12/9.

17. Daniel Searle to Council of State, August 28, 1653, P.R.O., C.O. 1/12/9; Petition of London Merchants, January 14, 1654, P.R.O., C.O. 1/12/6.I.

18. Daniel Searle to Council of State, September 19, 1653, P.R.O., C.O. 1/12/12; Daniel Searle to Council of State, October 19, 1653, P.R.O., C.O. 1/12/13.

19. Daniel Searle to Council of State, September 19, 1653, P.R.O., C.O. 1/12/12.

20. Memorial of the Assembly of Barbados to the Governor and Council, P.R.O., C.O. 1/12/12.I.

21. Memorial of the Assembly of Barbados to the Governor and Council, P.R.O., C.O. 1/12/12.I.

22. Petition of the Representative Body of Freeholders in Barbados to Cromwell, P.R.O., C.O. 1/12/12.II.

23. Daniel Searle to Council of State, September 19, 1653, P.R.O., C.O. 1/12/12; Daniel Searle to Council of State, October 19, 1653, P.R.O., C.O. 1/12/13.

24. Daniel Searle to Council of State, September 19, 1653, P.R.O., C.O. 1/12/12; Daniel Searle to Council of State, October 19, 1653, P.R.O., C.O. 1/12/13.

25. Daniel Searle to Council of State, September 19, 1653; P.R.O., C.O. 1/12/12; Daniel Searle to Council of State, October 19, 1653, P.R.O., C.O. 1/12/13.

26. An Estimate of the Barbados and of the Now Inhabitants There, Egerton Ms. 2395 (British Library), fol. 625.

27. Petition of Merchants of London, January 14, 1654, P.R.O., C.O. 1/12/16.16.II.

28. A Petition of the Council of the Island of Barbados, March 1654, John Thurloe, *A Collection of the State Papers of John Thurloe*, ed. Thomas Birch (London, 1742), p. 200; Daniel Searle to Oliver Cromwell, Lord Protector, March 30, 1654, ibid., pp. 199–200.

29. S. A. G. Taylor, *The Western Design* (Kingston, Jamaica: Institute of Jamaica and the Jamaican Historical Society, 1965); Sir Alan Burns, *History of the British West Indies* (London: George Allen and Unwin, 1954), pp. 246–61.

30. A paper of Col. Muddiford concerning the West Indies, Thurloe, *State Papers*, vol. 3, pp. 62–63; Vincent T. Harlow, *A History of Barbados, 1625–1685* (Oxford: Clarendon Press, 1926), p. 105.

31. Journal of West Indies Expedition, WYN/10/2 (National Maritime Museum, London); General Penn's Journal, WYN/10 (National Maritime Museum, London).

32. Thomas Modyford to Charles Modyford, Thurloe, *State Papers*, vol. 3, pp. 556–67; Harlow, *History of Barbados*, p. 107.

33. J. Berkehead to John Thurloe, Thurloe, *State Papers*, vol. 3, pp. 157–59; Thomas

Modyford to Charles Modyford, Thurloe, *State Papers*, vol. 3, pp. 566–67; Harlow, *History of Barbados*, pp. 107–8.

34. Thomas Modyford to Charles Modyford, Thurloe, *State Papers*, vol. 3, pp. 566–67.

35. Lieutenant-Colonel Francis Barrington to Sir John Barrington, June 6, 1655, Historical Manuscript Commission, seventh report, p. 572.

36. "The Narrative of General Venables," ed. C. H. Firth (London: Royal Historical Society, 1900), p. 12.

37. Lieutenant-Colonel Francis Barrington to Sir John Barrington, June 6, 1655, Historical Manuscript Commission, seventh report, p. 572.

38. Harlow, *History of Barbados*, pp. 110–11.

39. Thomas Modyford to Charles Modyford, Thurloe, *State Papers*, vol. 3, pp. 566–67.

40. Martin Noel to Governor Daniel Searle, August 27, 1657, Additional Ms. 11411 (British Library), fol. 45; Thomas Povey to William Povey, August 20, 1657, Additional Ms. 11411 (British Library), fol. 105; Thomas Povey to Daniel Searle, August 27, 1657, Additional Ms. 11411 (British Library), fols. 106–9; Henry Coventry to Jonathan Atkins, November 28, 1676, Additional Ms. 25,120 (British Library), fols. 96–99, 120.

41. The Humble Petition of Francis Craddock, Provost Marshall General of Barbados, February 8, 1669, P.R.O., C.O. 1/25/8; Reasons of the Council of Barbados Humbly Representing the Inconveniences which may Happen to This Island by Execution of a Grant of the Provost Marshall Place to Francis Craddock Esq. During His Natural Life by Patent Under the Great Seal, P.R.O., C.O. 1/18/77; Henry Coventry to Jonathan Atkins, November 28, 1676, Additional Ms. 25,120 (British Library), fols. 96–99.

42. The secretary's commission, Davis Collection (Roy. Com. Soc.), box 4, envelope 15; Richard Hall, *Acts, Passed in the Island of Barbados from 1643 to 1762* (London, 1764), pp. 8–11.

43. Charles Andrews, *British Committees, Commissions, and Councils of Trade and Plantations, 1627–1675*, John Hopkins University Studies in Historical and Political Science, Series 26 (Baltimore: The Johns Hopkins Press, 1908), pp. 24–60.

44. Thomas Povey to William Povey, November 10, 1655, Additional Ms. 11411 (British Library), fol. 15 and passim.

45. Thomas Povey to Daniel Searle, August 27, 1657, Additional Ms. 11411 (British Library), fols. 106–9.

46. Thomas Povey to William Povey, n.d., Additional Ms. 11411 (British Library), fols. 13–14.

47. Thomas Povey to Daniel Searle, January 8, 1657, Additional Ms. 11411 (British Library), fols. 123–27.

48. Thomas Povey to Daniel Searle, August 27, 1653, Additional Ms. 11411 (British Library), fols. 106–9; Thomas Povey to Daniel Searle, January 8, 1657–58, Additional Ms. 11411 (British Library), fols. 123–27; Thomas Povey to Daniel Searle, January 9, 1657–58, Additional Ms. 11411 (British Library), fol. 127; Thomas Povey to Daniel Searle, March 27, 1658, Additional Ms. 11411 (British Library), fols. 133–37; Daniel Searle to Martin Noel, May 10, 1658, Additional Ms. 11411 (British Library), fols. 150–60; Harlow, *History of Barbados*, pp. 119–20.

49. Thomas Povey and Martin Noel to Daniel Searle, April 30, 1659, Additional Ms. 11411 (British Library), fols. 184–86.

50. Thomas Povey to Daniel Searle, Additional Ms. 11411 (British Library), fol. 187;

Thomas Povey to Governor Searle, October 20, 1659, Egerton Ms. 2395 (British Library), fol. 176; Harlow, *History of Barbados*, pp. 122–23.

51. To the Supreame Authority the Parliament of the Commonwealth of England. The Humble Petition of the Representatives of the Island of Barbados for and in Behalf of the Inhabitants Thereof, Egerton Ms. 2395 (British Library), fol. 182; Harlow, *History of Barbados*, pp. 124–25.

52. Minutes of the Council of Barbados, July 31, 1660, P.R.O., C.O. 31/1; Thomas Modyford's Commission to be Governor of Barbados, P.R.O., C.O. 31/1/13–14; Harlow, *History of Barbados*, p. 125.

53. Harlow, *History of Barbados*, p. 126; Burns, *History of the British West Indies*, p. 297.

54. The most detailed look at efforts to bring proprietary government back to Barbados in the Restoration period is to be found in James Williamson, *The Caribbee Islands under the Proprietary Patents* (Oxford: Oxford University Press, 1926), pp. 21–63, 198–217. Williamson takes the position that Sir William Courteen, a member of the Powell Company, had some legal claim to the island, and his view of Restoration politics in Barbados is somewhat distorted by this opinion. For some modification of Williamson see Gary Puckrein, "Did Sir William Courteen Really Own Barbados?" *Huntington Library Quarterly* 44 (1981):136–37; Burns, *History of the British West Indies*, pp. 297–98; Harlow, *History of Barbados*, pp. 128–30.

55. Burns, *History of the British West Indies*, pp. 298–99; Harlow, *History of Barbados*, pp. 130–31.

56. Burns, *History of the British West Indies*, p. 299.

57. Ibid., pp. 299–300; Harlow, *History of Barbados*, pp. 132–44.

58. Planters were still complaining about the duty at the end of the eighteenth century. See Bryan Edwards, *The History, Civil and Commercial, of the British Colonies in the West Indies*, vol. 1 (London, 1801), pp. 508–9, 518–21; Burns, *History of the British West Indies*, p. 303; Harlow, *History of Barbados*, pp. 155–56.

59. Burns, *History of the British West Indies*, pp. 303–4; Harlow, *History of Barbados*, pp. 155–66.

60. Harlow, *History of Barbados*, p. 163; Burns, *History of the British West Indies*, pp. 304–5.

61. Burns, *History of the British West Indies*, pp. 308, 311.

62. Governor William Willoughby to King, December 8, 1666, P.R.O., C.O. 1/20/194.

63. Jill Sheppard, *The "Redlegs" of Barbados* (New York: KTO Press, 1977), p. 58.

64. William Willougby to Privy Council, December 16, 1667, P.R.O., C.O. 1/21/162.

65. The Humble Address and Petition of the Representatives of the Island of Themselves and the Inhabitants Thereof, September 5, 1667, P.R.O., C.O. 1/20/102.

66. Observations to the Barbados Petition [January 1668], P.R.O., C.O. 1/22/23.

67. Governor William Willoughby to King, December 8, 1666, P.R.O., C.O. 1/20/194.

68. Sir Peter Colleton and Other Planters in London to Christopher Cordrington, Deputy Governor, the Council, and Assembly of Barbados, December 14, 1670, *C.S.P.C. (1669–1674)*, p. 141; The Gentlemen Planters in London to the Assembly of Barbados, February 17, 1671, ibid., pp. 162–63.

69. Acts of Barbados, P.R.O., C.O. 30/2/89.

70. Acts of Barbados, P.R.O., C.O. 30/2/89.

71. The Assembly of Barbados to the Gentlemen Planters in London, June 16, 1671, *C.S.P.C. (1669–1674)*, p. 230.

72. Acts of Barbados, P.R.O., C.O. 30/2/272.

73. Henry Coventry to Sir Jonathan Atkins, November 28, 1676, Additional Ms. 25120 (British Library), fols. 96–97.

74. Henry Coventry to Sir Jonathan Atkins, November 28, 1676, Additional Ms. 25120 (British Library), fols. 96–99.

CHAPTER 9

1. Marcus Jernegan, *Laboring and Dependent Classes in Colonial America, 1607–1783* (Westport, Conn.: Greenwood Press, reprinted 1980), pp. 24–44.

2. The census is filed in P.R.O., C.O. 1/44/142–379.

3. Dunn, *Sugar and Slaves*, pp. 106–10.

4. William Bates had a plantation in St. Michael and a house in Bridgetown, P.R.O., C.O. 1/44/171.

5. P.R.O., C.O. 1/44/145, 146; A Coppie Journall Entries Made in the Custom House of Barbados Beginning August the 10th, 1664 and Ending August the 10th, 1665, Ms. Eng. Hist. b. 122 (Bodleian Library), passim; Wilfred Samuel, *The Jewish Colonists in Barbados in the Year 1680* (London: Jewish Historical Society of England, 1936), pp. 17–20.

6. Jerome S. Handler and Ron Shelby, "A Seventeenth-Century Commentary on Labor and Military Problems in Barbados," *Jour. Bar. Mus. Hist. Soc.* 34 (1973): 121.

7. Russell R. Menard, "The Maryland Slave Population, 1658 to 1730: A Demographic Profile of Blacks in Four Countries," *William and Mary Quarterly*, 3d ser., 32 (1975): 29–54; Ira Berlin, "Time, Space, and the Evolution of Afro-American Society in British Mainland North America," *American Historical Review* 85 (1980): 44–78.

8. Edward Littleton, *Groans of the Plantations* (London, 1689), pp. 6, 18–20.

9. *Memorials of the Great Civil War in England from 1646 to 1652*, ed. Henry Cary, vol. 2 (London: H. Colburn, 1842), p. 313.

10. Recopied Will Book, Philip Bell, December 9, 1658, Bar. Archives, RB 6/14/344.

11. *Barbados Records, Wills and Administrations, 1639–1680*, comp. and ed. by Joanne Mcree Sanders (Missouri: Sanders Historical Publications, 1979), p. 1.

12. Bridenbaugh, *No Peace*, pp. 120–21.

13. P.R.O., C.O. 1/44/234, 235.

14. J. Harry Bennett, *Bondsmen and Bishops*, University of California Publications in History, no. 62 (Berkeley: University of California Press, 1958), pp. 63–74; Frank Pitman, "Slavery in the British West India Plantations in the Eighteenth Century," *Journal of Negro History* 11 (1926): 605; Ulrich Philips, "An Antigua Plantation, 1769–1818," *North Carolina Historical Review* 3 (1926): 439–45.

15. Bennett, *Bondsmen and Bishops*, pp. 68–89.

16. P.R.O., C.O. 1/44/176, 241.

17. P.R.O., C.O. 1/44/198.

18. P.R.O., C.O. 1/44/173.

19. P.R.O., C.O. 1/44/230.

20. P.R.O., C.O. 1/44/178.

21. Berlin, "Time, Space, and the Evolution of Afro-American Society," pp. 50–51.

22. Dunn, *Sugar and Slaves*, p. 313.

23. Philip Curtin, *The Atlantic Slave Trade: A Census* (Madison: The University of Wisconsin Press, 1969), pp. 52–64, 72–73, 119, 216; Dunn, *Sugar and Slaves*, p. 314.

24. Dunn, *Sugar and Slaves*, p. 314.

25. K. G. Davies, *The Royal African Company* (New York: Longmans, Green, 1957), pp. 299–300.

26. Dunn, *Sugar and Slaves*, p. 315.

27. Recopied Deed Books, Bar. Archives, RB 3/17/263.

28. Recopied Deed Books, Bar. Archives, RB 3/6/136.

29. Dunn, *Sugar and Slaves*, p. 317.

30. Bennett, *Bondsmen and Bishops*, p. 53.

31. Ibid., pp. 53–56.

32. Recopied Deed Books, Bar. Archives, RB 3/13/132.

33. Littleton, *Groans of the Plantations*, pp. 6, 18.

34. Instructions I would have observed by Mr. Richard Harwood in the management of my plantation, Rawlinson Ms. A 348 (Bodleian Library), fol. 3.

35. Ibid., fol. 4.

36. Richard Ligon, *A True and Exact History of Barbados* (London, 1657), p. 47.

37. Monica Schuler argues for the retentions of African ethnic identities among new Negroes, but the Barbadian evidence would seem to suggest differently. See her "Afro-American Culture," *Historical Reflections* 6 (1979):121–37. Richard Price's comments on her essay in the same volume should also be read.

38. B. W. Higman, "African and Creole Slave Family in Trinidad," *Journal of Family History* 3 (1978):172–73.

39. Governor Sir Jonathan Atkins to Secretary Sir Joseph Williamson, October 3, 1675, *C.S.P.C. (1675–1676)*, pp. 294–95.

40. Bennett, *Bondsmen and Bishops*, p. 34.

41. Jerome S. Handler, "Slave Insurrectionary Attempts in Seventeenth-Century Barbados" (Thirteenth Conference of the Association of Caribbean Historians, Pointe-à-Pitre, Guadeloupe, 1981), p. 21.

42. R. S. Rattray, *Ashanti* (London: Oxford University Press, 1923), pp. 288–93.

43. Handler, "Slave Insurrectionary Attempts," p. 21.

44. Richard Hall, *Acts, Passed in the Island of Barbados from 1643 to 1762* (London, 1764), p. 472.

45. Ligon, *Barbados*, p. 46.

46. Governor William Willoughby to Lords of the Council, July 9, 1668, *C.S.P.C. (1661–1668)*, p. 586.

47. Handler, "Slave Insurrectionary Attempts," p. 40.

48. Ibid., p. 46.

49. Ibid.

50. Morgan Godwin, *The Negro's and Indian's Advocate* (London, 1680), p. 111.

51. Jernegan, *Laboring and Dependent Classes*, pp. 24–44.

52. Minutes of the Barbadian Council, Transcript, 1689–96, Bar. Archives, fols. 215–16.

53. Godwin, *The Negro's Advocate*, pp. 38–39.

54. Ibid., p. 39.

55. "Extracts from Henry Whistler's Journal of the West Indian Expedition," in *Narrative of General Venables with an Appendix of Papers Relating to the Expedition to the West Indies and the Conquest of Jamaica*, ed. C. H. Firth (London: Longmans, Green, 1900), p. 146.

56. Thomas Jefferson, *Notes on the State of Virginia*, ed. William Peden (Chapel Hill: University of North Carolina Press, Institute of Early American History and Culture, 1955), p. 162.

57. Acts of Barbados, P.R.O., C.O. 30/2/16.

58. Ligon, *Barbados*, p. 46.

59. Godwin, *The Negro's Advocate*, passim; Thomas Tryon, *Friendly Advice to Gentlemen Planters of East and West Indies* (London, 1684), pp. 75–222.

60. Winthrop Jordan, *White Over Black* (Chapel Hill: University of North Carolina Press, Institute of Early American History and Culture, 1968), pp. 484–85, 491–93.

61. Minutes of the Assembly, March 21, 1675, Bar. Archives.

CHAPTER 10

1. Edmund Morgan, *American Slavery, American Freedom* (New York: W. W. Norton, 1975), pp. 338–87, esp. pp. 344–45.

2. Donald Robinson, *Slavery in the Structure of American Politics, 1765–1820* (New York: Harcourt Brace Jovanovich, 1971), p. 98.

3. Ibid., p. 99.

4. Ibid.

5. Ibid., p. 105.

6. Benjamin Quarles, *The Negro in the American Revolution* (Chapel Hill: University of North Carolina Press, 1961), pp. 19–32.

7. Quoted in George Washington Williams, *History of the Negro Race in America*, vol. 1 (New York: G. P. Putman's Sons, 1883), p. 341.

8. Robinson, *Slavery*, pp. 109–10, 128–30.

9. George Livermore, *An Historical Research Respecting the Opinions of the Founders of the Republic on Negroes as Slaves, as Citizens, and as Soldiers* (Boston: A. Williams, 1863), p. 56. Actually, the instructions to the commissioners mentioned only the freeing of servants. See "Virginia in 1650–52," *The Virginia Magazine of History and Biography* 17 (1809): 283.

10. Morgan, *American Slavery*, pp. 11–12.

11. David W. Cohen and Jack Greene, *Neither Slave nor Free* (Baltimore: The Johns Hopkins University Press, 1972), pp. 13–17.

12. C. L. R. James, *The Black Jacobins* (New York: Vintage Books, 1963).

13. John Lombardi, *The Decline and Abolition of Negro Slavery in Venezuela* (Westport, Conn.: Greenwood Press, 1971), pp. 35–53 and passim.

14. Arthur Corwin, *Spain and the Abolition of Slavery in Cuba, 1817–1886* (Austin: University of Texas Press, Institute of Latin American Studies, 1967), pp. 293–313.

APPENDIX

1. Frank Pitman, *The Development of the British West Indies, 1700–1763* (New Haven: Yale University Press, 1917), pp. 369–70.

2. Richard Dunn, *Sugar and Slaves*, pp. 62, 67, 75, 88–89, 300–301, 334, 337, 340.

3. *The Record of the Virginia Company of London*, ed. Susan M. Kingsbury, vol. 4 (Washington, D.C.: United States Government Printing Office, 1935), p. 58; Alexander Brown, *The First Republic in America* (New York: Houghton Mifflin, 1898), passim.

4. St. Julien Childs, *Malaria and Colonization in the Carolina Low Country*, The Johns Hopkins University Studies in Historical and Political Science 58, no. 3 (Baltimore: The Johns Hopkins Press, 1940), pp. 218–64; Peter Wood, *Black Majority: Negroes in Colonial South Carolina* (New York: Knopf, 1974), pp. 63–91.

5. Richard Ligon, *A True and Exact History of Barbados* (London, 1657), p. 27.

6. L. Schuyler Fonnaroff, "Did Barbados Import Its Malaria Epidemic?" *Jour. Bar. Mus. Hist. Soc.* 34 (1973): 122–30.

7. Robert Schomburgk, *The History of Barbados* (London: Frank Cass, reprinted 1971), pp. 73–74.

8. *Winthrop Papers*, vol. 5, pp. 219–20.

9. *Winthrop's Journal*, ed. James Hosmer, *Original Narratives of Early American History*, vol. 2 (New York: Scribner's, 1908), p. 329.

10. Ligon, *Barbados*, p. 25.

11. J. H. Bennett, "Peter Hay, Proprietary Agent in Barbados," *Jamaican Historical Review* 5 (1965):12–13.

12. Peter Hay to Archibald and James Hay, August 22, 1640, Hay Papers (S.R.O.).

13. *Winthrop's Journal*, vol. 2, p. 328.

14. Nicholas Blake to Joseph Williamson, March 23, 1670, *C.S.P.C.* (*1675–1676*), p. 59.

15. Otis Starkey, *The Economic Geography of Barbados* (New York: Columbia University Press, 1939), pp. 8–9.

16. Governor Jonathan Atkins to Secretary Sir Joseph Williamson, October 3, 1675, *C.S.P.C.* (*1675–1676*), p. 294.

17. Governor Jonathan Atkins to Secretary Sir Joseph Williamson, October 3, 1675, ibid., p. 294.

18. Governor Lord Vaughan to Secretary Sir Joseph Williamson, September 20, 1675, ibid., p. 282.

19. Secretary Sir Joseph Williamson to Lord Vaughan, Governor of Jamaica, December 6, 1675, ibid., p. 311.

20. Dunn, *Sugar and Slaves*, pp. 88–89, 108–10.

21. Patricia Molen, "Population and Social Patterns in Barbados in the Early Eighteenth Century," *William and Mary Quarterly*, 3d ser., 28 (1971): 287–300.

22. Dunn, *Sugar and Slaves*, pp. 325–32.

23. Ibid., p. 109, fn. 36.

24. James Camden Hotten, *List of Emigrants to America, 1600–1700* (New York: Bouton, 1874), p. 422.

25. Ibid., pp. 490–91.

26. Dunn, *Sugar and Slaves*, p. 328.

27. Governor Jonathan Atkins to Lords of Trade and Plantations, March 26, 1680, P.R.O., C.O. 1/44/45; Richard Dutton to Lords of Trade and Plantations, August 19, 1681, P.R.O., C.O. 1/47/7.

28. See below, pages 192–193.

29. *Daily Journal of Major George Washington*, ed. J. M. Toner (New York: Joel Munsell's Sons, 1892), pp. 40–41; An Abstract of all Persons, Born, Christened, and Buried in this Island, P.R.O., C.O. 28/32/F.f. 26.

30. Gerald Grob, "Disease and Environment in American History," *Handbook of Health, Health Care and the Health Professions*, ed. David Mechanic (New York: The Free Press, forthcoming); Billy Smith, "Death and Life in a Colonial Immigrant City: A Demographic Analysis of Philadelphia," *Journal of Economic History* 37 (1977): 871.

31. Africa was not uniformly unhealthy for Europeans. Feinberg has found that the Dutch on the Gold Coast of Africa did not experience a high mortality rate. H. M. Feinberg, "New Data on European Mortality in West Africa, the Dutch on the Gold Coast, 1719–1760," *Journal of African History* 15 (1974): 357–71.

32. Baptized, Buried Abstract, January 1, 1783, to December 31, 1739, P.R.O., C.O. 28/25/A.a. 90.

33. *Massachusetts Historical Society, Collections*, 5th ser., vol. 8, p. 384.

34. Governor Sir Jonathan Atkins to Lords of Trade and Plantations, October 26, 1680, *C.S.P.C.* (*1677–1680*), p. 619.

35. Deputy Governor Stede to Lords of Trade and Plantations, October 17, 1685, *C.S.P.C.* (*1685–1688*), p. 109; Journal of Assembly of Barbados, November 17, 1685, ibid., p. 121; Deputy Governor Stede to Lords of Trade and Plantations, January 8, 1685/86, ibid., p. 139.

36. John Oldmixon, *The British Empire in America*, vol. 2 (London, 1708), p. 102.

37. Governor Russell to Lords of Trade and Plantations, September 25, 1694, *C.S.P.C.* (*1693–1696*), p. 362.

38. Minutes of the General Assembly of Barbados, May 7, 1700, *C.S.P.C.* (*1700*), p. 239.

39. Governor Sir Bevill Granville to Council of Trade and Plantations, June 16, 1703, *C.S.P.C.* (*1702–1703*), p. 503.

40. Pitman, *British West Indies*, pp. 372–73; Dunn, *Sugar and Slaves*, p. 87; A List of Inhabitants, P.R.O., C.O. 28/14/2.v.

41. Governor Mitford Crowe to Council of Trade and Plantations, September 2, 1709, *C.S.P.C.* (*1708–1709*), p. 457.

42. Oldmixon, *British Empire*, vol. 2, p. 112.

43. James Hendy, *A Treatise on the Glandular Disease of Barbados* (London, 1784), p. 33.

44. See, for example, Richard Towne, *A Treatise of the Diseases Most Frequent in the West Indies, and Herein More Particularly of Those which Occur in Barbados* (London, 1726); Henry Warren, *A Treatise Concerning the Malignant Fever in Barbados* (London, 1741); William Hillary, *Observations on the Changes of the Air and the Concomitant Epidemical Disease of the Island of Barbados and Other West Indian Islands* (London, 1754).

Bibliographic Essay

The notes to this monograph provide the documentation to support the ideas expressed here. This bibliographic essay is intended as a supplementary guide to the works that were found to be most useful in the preparation of this study. It is by no means an exhaustive list, and certainly the absence of a particular work should not be interpreted as a comment about its intrinsic worth.

Fortunately, just as I was beginning my research four major bibliographic compilations were published: Michael J. Chandler, *A Guide to Records in Barbados* (Oxford: Basil Blackwell, 1965); Jerome S. Handler, *Guide to Materials for Study of Barbados History, 1627–1834* (Carbondale: Southern Illinois University Press, 1971); Kenneth Ingram, *Manuscripts Relating to Commonwealth Caribbean Countries in United States and Canadian Repositories* (London: Caribbean University Press, 1971); Peter Walne, *A Guide to Manuscript Sources for the History of Latin America and the Caribbean in the British Isles* (London: Oxford University Press, 1973). I am deeply indebted to these authors, and I strongly recommend their works to anyone interested in doing historical research on Barbados or the larger English-speaking Caribbean.

There are several useful introductory works to early Caribbean history: Arthur Newton, *The European Nations in the West Indies, 1493–1688*

(London: A. and C. Black, 1937), lacks footnotes but is a well-written narrative that gets the basic chronology right; Arthur Newton, *The Colonising Activities of English Puritans* (New Haven: Yale University Press, 1914), is still necessary reading, though it is a bit thin on the social background. Sir Alan Burns's *History of the British West Indies* (London: George Allen and Unwin, 1954) is a first-rate work, especially for the seventeenth century. The recent anthology, *The Westward Enterprise*, edited by K. R. Andrews, N. P. Canny, and P. E. Hair (Detroit: Wayne State University Press, 1979), has several fine essays that shed light on the dark period of 1590–1624. The works of Carl and Roberta Bridenbaugh, *No Peace Beyond the Line* (New York: Oxford University Press, 1972); Richard Dunn, *Sugar and Slaves* (Chapel Hill: University of North Carolina Press, Institute of Early American History and Culture, 1972); and Richard Sheridan, *Sugar and Slavery* (Baltimore: The Johns Hopkins University Press, 1973), provide the social and economic background that the earlier political works lacked. Although primarily concerned with the eighteenth century, also useful are Frank Pitman, *The Development of the British West Indies, 1700–1763* (New Haven: Yale University Press, 1917), and Lowell Ragatz, *The Fall of the Planter Class in the British Caribbean, 1763–1833* (New Haven: The Century Co., 1928).

James Williamson has done the standard work on the first settlement of Barbados. His study is primarily concerned with a rivalry between the earl of Carlisle and Sir William Courteen for proprietary control of the island and pays little attention to social developments in the colony. His Carlisle/Courteen thesis has been challenged. See Gary Puckrein, "The Carlisle Papers," *Jour. Bar. Mus. Hist. Soc.* 35 (1978): 300–315, and "Did Sir William Courteen Really Own Barbados?" *Huntington Library Quarterly* 44 (1981):135–57.

Besides the works of Bridenbaugh, Dunn, and Sheridan, those interested in the social and economic development of presugar Barbados should also consult Otis Starkey, *The Economic Geography of Barbados* (New York: Columbia University Press, 1939), and David Watts, *Man's Influence on the Vegetation of Barbados, 1627 to 1800*, Occasional Papers in Geography, no. 4 (Hull, England: University of Hull, 1966); Richard Pares, "Merchants and Planters," Economic History Review Supplements, no. 4 (Cambridge: Cambridge University Press, 1960); and F. C. Innes, "The Pre-Sugar Era of European Settlement in Barbados," *Journal of Caribbean History* 1 (1970):1–22.

The political history of seventeenth-century Barbados is dominated by the older and flawed works of Vincent Harlow, Nicholas Davis (see Preface), and James Williamson. J. H. Bennett, "Peter Hay, Proprietary Agent in Barbados," *Jamaican Historical Review* 5 (1965): 9–29, and "The English Caribbees in the Period of the Civil War, 1642–1646," *William and Mary Quarterly*, 3d ser., 24 (1967):359–77, are the best of modern scholarship. Bennett relies almost completely upon the Hay of Haystoun Papers (S.R.O.), and his work needs to be supplemented with other source material, particularly from the British Public Record Office.

Two alternative interpretations of the coming of sugar cultivation to Barbados are Matthew Edels, "The Brazilian Sugar Cycle of the Seventeenth Century and the Rise of the West Indian Competition," *Caribbean Studies* 9 (1969): 24–44, and Robert Carlyle Bate, "Why Sugar? Economic Cycles and the Changing of Staples in the English and French Antilles, 1624–1654," *Journal of Caribbean History* 8 (1976): 1–41.

The literature on the origins of Negro slavery is enormous and growing. For Barbados: Hilary Beckles, "The Economic Origins of Black Slavery in the British West Indies, 1640–1680: A Tentative Analysis of the Barbados Model," *Journal of Caribbean History* 16 (1982): 36–56, provides a Marxist approach (his chronology is wrong). An econometric treatment can be found in Richard Bean and Robert Thomas, "The Adoption of Slave Labor in British America," in *The Uncommon Market*, edited by Henry A. Gemery and Jan S. Hogendorn (New York: Academic Press, 1975), pp. 377–98. A sampling of other theories can be found in the excellent anthology edited by Donald Noel, *The Origins of American Slavery and Racism* (Columbus, Ohio: Charles E. Merrill, 1972).

The life of poor whites and indentured servants has been variously treated by Hilary Beckles, "Rebels and Reactionaries: The Political Responses of White Labourers to Planter-class Hegemony in Seventeenth-Century Barbados," *Journal of Caribbean History* 15 (1981):1–19; "Land Distribution and Class Formation in Seventeenth-Century Barbados: The Rise of a Wage Proletariat," *Jour. Bar. Mus. Hist. Soc.* 36 (1980):136–44; "Sugar and Servitude: An Analysis of Indentured Labour during the Sugar Revolution of Barbados, 1643–1655," *Jour. Bar. Mus. Hist. Soc.* 36 (1981): 236–47; and by Jill Sheppard, *The "Redlegs" of Barbados* (New York: KTO Press, 1977). Beckles is a bright and creative scholar, but his ideological proclivities frequently cause him to draw conclusions that his evidence cannot support.

We still wait for a prosopographical study of the Barbadian planter class. Richard Waterhouse, "England, the Caribbean, and the Settlement of Carolina," *Journal of American Studies* 9 (1975): 259–81, offers some preliminary remarks. The publication of abstracts of Barbadian wills that have survived in the Barbadian Department of Archives, *Barbados Records, Wills and Administrations, 1639–1680*, compiled and edited by Joanne Mcree Sanders (Missouri: Sanders Historical Publications, 1979), have set the stage for such a study. The *Jour. Bar. Mus. Hist. Soc.* contains several genealogical studies of planter families.

There is no treatment of a Barbadian plantation such as Michael Craton and James Walvin, *A Jamaican Plantation: The History of Worthy Park 1670–1970* (London: W. H. Allen, 1970), though J. H. Bennett, *Bondsmen and Bishops*, University of California Publications in History, no. 62 (Berkeley: University of California Press, 1958), comes close.

The emergence of plantation culture has been a subject of wide discussion with most attention focused on the study of continuities or changes in African cultural traditions in the New World. A useful survey of the literature can be found in Frederick Cooper, *Plantation Slavery on the East Coast of Africa* (New Haven: Yale University Press, 1977). Jerome S. Handler and Frederick Lange, *Plantation Slavery in Barbados* (Cambridge: Harvard University Press, 1978), have conducted excavations of slave graves on Newton Plantation, and they have made some worthwhile observations about the cultural emergences of an Afro-Barbadian population on the island. Monica Schuler, "Afro-American Culture," *Historical Reflections* 6 (1979):121–37, is a brilliant look at the evolution of a new culture among Jamaican slaves. The commentaries on her essay by Mary Karasch, Richard Price, and Edward Kamau Brathwaite in the same volume should also be read with care. Edward Brathwaite, *The Development of Creole Society in Jamaica* (Oxford: Clarendon Press, 1971), urges us to consider the changes that European social patterns underwent in a plantation economy.

The best work on slave revolts in Barbados is Jerome S. Handler, "Slave Insurrectionary Attempts in Seventeenth-Century Barbados" (Thirteenth Conference of the Association of Caribbean Historians, Pointe-à-Pitre, Guadeloupe, 1981), though it lacks an interpretive framework, and thus should be supplemented with the studies of Michael Craton, "The Passion to Exist: Slave Rebellions in the British West Indies 1650–1832," *Journal of Caribbean History* 13 (1980):1–20; Anthony Synott, "Slave Revolts in the Caribbean" (Ph.D. diss., University of London, 1976); Monica Schuler, "Akan Slave Rebellions in the

British Caribbean," *Savacou* 1 (1970): 8–31; and David Barry Gaspar, "The Antigua Slave Conspiracy of 1736: A Case Study of the Origins of Collective Resistance," *William and Mary Quarterly*, 35 (1978): 308–23.

Index